Solid State Physics - Metastable, Spintronics Materials and Mechanics of Deformable Bodies - Recent Progress

Edited by Subbarayan Sivasankaran, Pramoda Kumar Nayak and Ezgi Günay

Published in London, United Kingdom

IntechOpen

Supporting open minds since 2005

Solid State Physics - Metastable, Spintronics Materials and Mechanics of Deformable Bodies - Recent Progress
http://dx.doi.org/10.5772/intechopen.78485
Edited by Subbarayan Sivasankaran, Pramoda Kumar Nayak and Ezgi Günay

Contributors
Chih-Yuan Chen, Richard Oldani, Victor Lykah, Eugen Syrkin, Roberto Raúl Deza, Khalid Mahmood, Adnan Ali, Jolli Jacob, Sajad Hussain, Alina Manzoor, M. Imran Arshad, Tariq Munir, Sofia Akbar, M Ajaz Un Nabi, Salma Ikram, Hafiz Naeem Ur Rehman, Abdelaziz Bouchikhi, Iseli Nantes, Juliana Araujo-Chaves, David Lopes, Lucivaldo Menezes, Richard Pincak, Erik Bartos, Satya Pal Singh, Esakki Muthu Sankaran, Arumugam Sonachalam, Mohamed Abdelsabour Fahmy, Natela Zirakashvili, Xuewen Yin, Hui Li, Wenwei Wu

First published in London, United Kingdom, 2020 by IntechOpen
IntechOpen is the global imprint of INTECHOPEN LIMITED, registered in England and Wales, registration number: 11086078, 7th floor, 10 Lower Thames Street, London, EC3R 6AF, United Kingdom
Printed in Croatia

British Library Cataloguing-in-Publication Data
A catalogue record for this book is available from the British Library

Additional hard and PDF copies can be obtained from orders@intechopen.com

Solid State Physics - Metastable, Spintronics Materials and Mechanics of Deformable Bodies - Recent Progress
Edited by Subbarayan Sivasankaran, Pramoda Kumar Nayak and Ezgi Günay
p. cm.
Print ISBN 978-1-83881-164-8
Online ISBN 978-1-83881-165-5
eBook (PDF) ISBN 978-1-83881-166-2

For EU product safety concerns:
IN TECH d.o.o., Prolaz Marije Krucifikse Kozulić 3, 51000 Rijeka, Croatia, info@intechopen.com or visit our website at intechopen.com.

Meet the editors

Dr. S. Sivasankaran is working as an associate professor in the Department of Mechanical Engineering, College of Engineering, Qassim University. He has received several Deanship of Scientific Research Grants in Qassim University towards research work in the fields of materials and manufacturing. He completed his PhD degree in Production Engineering (Manufacturing Technology—Nanocomposite Materials) from NIT Trichy, India. After his doctoral degree, he worked as an associate professor in the KPR Institute of Engineering and Technology (KPRIET), Coimbatore, Tamilnadu, India. He has established an advanced materials research laboratory and entrepreneurship development cell in KPRIET. He has received a number of research and seminar grants from government of India. Furthermore, he served as an associate professor in Hawassa University, East Africa, for 2 years where he was involved in curriculum development. He has published more than 75 research articles in refereed international journals, acts as a reviewer for more than 20 refereed international journals (Elsevier, Springer, ASME), and has edited two books. Finally, his research interests are nanocomposites, mechanical alloying;, powder metallurgy, stir casting, advanced welding, 3D printing, metal matrix composites, polymer matrix composites, mechanical behavior, machining behavior, advanced characterization techniques, optimization, and modeling.

Pramoda Kumar Nayak is currently working as Adjunct Professor (Ramanujan Fellow) at the Department of Physics, Indian Institute of Technology, Madras, India. He received his doctoral degree in Physics from the Indian Institute of Technology, Guwahati, India. After completion of his PhD, he worked as a post-doctoral researcher and visiting scientist in several institutions around the globe, including the Institute for Plasma Research, India, National Cheng Kung University, Taiwan, National Tsing Hua University, Taiwan, and Ulsan National Institute of Science and Technology, Republic of Korea. His research interest is exploring intriguing physics in van der Waals materials and their heterostructures. He has published more than 50 international journals, five book chapters, one book, and three edited volumes so far in his research carrier.

Ezgi Günay graduated from the Engineering Sciences Department at the Middle East Technical University (METU, Ankara, Turkey) in 1985. She completed her Master of Science degree in 1989 at the same department. The title of her thesis was "Development of a Preprocessor and Modification of a Finite Element Procedure for the Analysis of Metal Forming Processes" (December 1989, METU). She received her PhD from the Mechanical Engineering Department at Gazi University in 1996. The title of the thesis was "A Nonlocking Finite Element Model for Nonlinear Analysis of Thin and Thick

Composite Plates." She worked as an assistant professor between 1999 and 2009 and studied academically by giving basic courses as an associate professor between 2010 and 2020 at the same department. During these years, she gave courses on the following subjects: technical drawing, FORTRAN-computer programming languages, applied mathematics for mechanical engineers, differential equations, statics, dynamics, strength of materials, introduction to numerical analysis, introduction to composite materials, introduction to finite element analysis, finite element method, plate and shell theories, and elasticity. She has authored about 40 papers published both in national and international proceedings and journals. She has written two books and has had three chapters published in international books.

Contents

Preface

In modern materials science, solid-state physics is a multidisciplinary field that describes the advancements in physics, chemistry, and material engineering. Solid-state physics deals with the behavior of solid matters in terms of physical metallurgy, quantum mechanism, crystallography, and electromagnetism. The concept of solid-state physics is currently being applied to all electronic parts, which is a boon for the electronics industry. The parts made up of solid matters are categorized as crystalline solids and amorphous materials/bulk metallic glasses/non-equilibrium materials. Solid-state materials exhibit interesting properties: higher strength, hardness, increased elastic strain limit, and outstanding corrosion and wear resistance. In addition, solid-state materials usually have unique physical, thermal, magnetic, and electrical properties, which are triggered to apply these materials to structural or functional applications. Furthermore, these materials possess macroscopically homogeneous, isotropic, and superior plastic deformation abilities. Therefore, to investigate and demonstrate the field of solid-state physics, this book addresses recent progress in the field of solid-state physics, which includes scientific works and reviews related to metastable and spintronics materials.

Dr. Subbarayan Sivasankaran
College of Engineering,
Qassim University,
Saudi Arabia

"Spintronics" usually refers to the branch of physics concerned with the manipulation, storage, and transfer of information by means of electron spins in addition to the electron charge in conventional electronics. It is very important to understand the principles and equations underlying the physics, transport, and dynamics of spin in solid-state systems. Major advances in electron spin transport started with the discovery of room temperature giant magnetoresistance, which paves the way towards application in spin-based practical devices such as spintronic field-effect transistors, spin-dependent tunneling diodes, logic gates, quantum computers, etc. The study of spintronics in semiconductors, metals, and other materials has been widely explored in its bulk form. The recent emergence of two-dimensional (2D) materials has been a real boom in the field of spintronics due to the strong spin–orbit coupling effect. The aim of the "Spintronics" section is to provide recent development in spintronics in bulk as well as 2D materials aimed at researchers, professors, post-doctorates, and graduate students in the discipline of physics, materials science, and nanotechnology and to help them master the overall knowledge of spintronics.

The section contains three chapters. In Chapter 1, the authors discuss spin—orbit coupling in an exotic graphene structure and also in biology. They introduce a new representation of the genetic code in the time series for string and D-brane modeling by applying a spinor field to a superspace in time series data. This method develops supersymmetry for living organisms, which is considered to be one of the big puzzles of modern biology. Chapter 2 introduces the Ising model and its applications, and highlights developments in the field of magnetism relevant to the field of spintronics. In Chapter 3, the authors report the effect of Ni/Mn variation on the exchange bias properties in Mn-rich Ni50 – xMn37 + xSn13 ($0 \leq x \leq 4$) Heusler alloys. The exchange bias properties in the above system are the key elements of new spintronics systems.

I am very pleased to serve as section editor of this book, which contains a wide variety of studies from different authors. I would like to thank all the authors for their effort in contributing their research papers.

I would also like to acknowledge the help given by IntechOpen, in particular publishing manager Lada Bozic for her assistance, patience, and support throughout the whole process of this book project.

Dr. Pramoda Kumar Nayak
Department of Physics,
IIT Madras,
Chennai, India

This aim of this section is to explain the newly developed numerical and analytical methods by describing their solutions. Additionally in this section, newly developed theoretical studies on the solution methods of basic problems for solid mechanics are included. Solid mechanics problems are discussed in two main groups: solutions of beam, plate, and shell-type structures under static loading and solutions under dynamic loading. To make the mathematical problems simpler and more understandable, the modeling is performed in terms of plane stress, plane strain, and axisymmetric and symmetric conditions. The section "Mechanics of Deformable Bodies" contains three chapters. The section includes the basic scientific knowledge that is required as well as newly developed analytical solution methods and numerical solution techniques that have novel commercial application areas in engineering field.

The first chapter of the section "2D Elastostatic Problems in Parabolic Coordinates" was prepared by Assoc. Prof. Dr. Natela Zirakashvili and Prof. Dr. I. Vekua. The researchers explain and discuss the boundary value problems that were considered in the defined parabolic coordinate system. In the parabolic coordinates, the equilibrium equations, Hooke's law, and analytical (exact) solutions of 2D problems of elasticity were constructed in the homogeneous isotropic body bounded by coordinate lines of the parabolic coordinate system. Analytical solutions were obtained using the method of separation of variables. The solution was constructed using its general representation by two harmonic functions. Using MATLAB software, numerical results and constructed graphs of some boundary value problems were obtained and presented in detail.

The second chapter, "Boundary element mathematical modelling and boundary element numerical techniques for optimization of micropolar thermoviscoelastic problems in solid deformable bodies," was prepared by Assoc. Prof. Dr. Mohamed Abdelsabour Fahmy. The main objective of this chapter was to introduce a new theory called three-temperature nonlinear generalized micropolar thermoviscoelasticity. Because of strong nonlinearity of simulation and optimization problems associated with this theory, the numerical solutions for problems related to the proposed theory are always very difficult and require the development of new numerical techniques. So, the researcher proposed a new boundary element technique for simulation and optimization of such problems based on genetic algorithm, freeform deformation method, and nonuniform rational B-spline curve as the shape optimization technique.

The third chapter, "Dynamic stiffness method for vibrations of ship structures," is by Dr. Xuewen Yin, researcher Kuikui Zhong, researcher Zitian Wei, and Prof. Dr. Wen-wei Wu. The chapter was prepared by considering the dynamic stiffness method (DSM) in the solutions of the dynamics of ship structures. A DSM element accounting for both in-plane and bending vibrations in flat rectangular plates was developed, which makes it possible for modeling wave conversion across junctions in built-up plates. In addition, a DSM element for stiffened plates was formulated, which considers all possible vibrations in plates and beams, i.e. bending, torsion, and extension motions. The third type of DS plate element was examined in terms of fluid loading, which was induced by a vibrating plate. Finally, the proposed DSM method was extended to address vibration transmission in a built-up plate structure, which demonstrated the great potential of DSM in applications of more practical and more general engineering fields.

Dr. Ezgi Günay
Associate Professor,
Engineering Faculty,
Mechanical Engineering Department,
Gazi University,
Ankara, Turkey

Section 1

Solid State Behavior

Chapter 1

Issues in Solid-State Physics

Roberto Raúl Deza

Abstract

In the first sections, we bring into the present context some of our past contributions on the influence of quantum correlations on the formation of tightly bound solids. We discuss the effects of the overlap between neighbor orbitals in diverse situations of interest—involving both bulk and surface states—and call the reader's attention to an exact tight-binding calculation which allows gauging the errors introduced by the underlying hypotheses of the usual tight-binding approximation. We round up this part by reviewing a quantum Monte Carlo method specific for strongly correlated fermion systems. In the last section, we explore some non-equilibrium routes to (not necessarily tightly bound) solid state: we discuss spatiotemporal pattern formation in arrays of FitzHugh-Nagumo (FHN) neurons, akin to resonant crystal structures.

Keywords: quantum correlations, band structure, tight-binding approach, neighbor orbital overlap, fermion Monte Carlo, non-equilibrium pattern formation, spatiotemporal synchronization

1. Introduction

Since childhood, we all have an intuition of what a solid is. However, most properties we intuitively assign to solids come in a vast range. Diamonds—and some metals—are hard, and ordinary glasses are brittle; but vulcanized rubber is neither, and it is a solid too. Perhaps the best characterization is this: *at our human timescales, a solid does not flow*. That is why this category includes glasses and ice (which do flow but at least at geological timescales).

Regarding their structure, a huge class of solids are *crystalline*. This is so to such extent that *solid state* came to be synonymous of crystalline structure, and the more comprehensive category of *condensed matter* (which admittedly includes condensed fluids or liquids) came into fashion. The name *crystal* was assigned in the late antiquity to precious and semiprecious stones that outstood for their transparency and diaphaneity. In fact, the modern meaning of the term as "an almost perfectly ordered structure" explains easily those properties.[1]

Many solids we interact with—metals, stones, etc.—are random assemblies of grains, held together by strong adhesion forces. Like those of sand, quartz, or salt, those grains are very likely to be themselves crystals (which as said do not imply they are perfect: they may contain lots of impurities and defects). But there are two particular aspects of crystals we are concerned with here. The first is that

---

[1] For isolators like these, the bandgap is too large for visible light to be absorbed by creating electron-hole pairs. Moreover, the absence of charge carriers rules out light scattering. Impurities provide localized midgap states, which favor two-step electron-hole pair creation by visible light.

unlike complex systems, which may display emergent structures at each scale (think, e.g., of mitochondria, cells, tissues, organs, etc.), crystals are very simple: they are huge assemblies of elementary building blocks (be they atoms, molecules, nanoclusters, or whatever). The second is that since the building blocks obey quantum mechanics, crystals inherit the quantum character (despite being themselves macroscopic).

As recent experiments have shown, whereas most interactions (but gravity) are effectively short-ranged, there is no limit for quantum correlations; and this fact makes them the most important fact to account for in modeling. Quantum correlations manifest themselves in many ways, but the by far dominant one comes from the indistinguishability of identical particles. Unless the crystal is a monolayer, the state vector of a system of many indistinguishable particles must be either *totally symmetric* or *totally antisymmetric* (a determinant) under exchange. In the first case, the particles obey Bose-Einstein statistics and are called *bosons*. In the second, the particles obey Fermi-Dirac statistics and are called *fermions*. The requirement that the state vector of a system with many fermions be totally antisymmetric is the celebrated *exclusion principle*, postulated by Pauli.

At present, there is no question that atoms are distinguishable. They can even be individually manipulated.[2] Since in modeling crystals, it suffices to take atoms as building blocks (we resolve up to the nanoscale), it does not matter that they are themselves composed of other indistinguishable particles (i.e., protons and neutrons, confined to $\sim 10^{-6}$ nm) besides electrons. Instead, considering the typical effective masses of electrons in metals and semiconductors, their thermal lengths at room temperature can reach the μm, so they are highly delocalized. The following two sections illustrate two different ways of dealing with Pauli's exclusion principle when modeling crystalline solids, corresponding to two radically different ways of doing quantum mechanics.

Section 2 keeps within the framework of *first quantization*. It is assumed that neither electrons (we mean *crystal* electrons, with effective masses) nor holes can be either created or destroyed. There is only one electron in the whole crystal, submitted to a potential which is mainly the juxtaposition of shielded Coulomb terms, due to atomic orbitals located at the crystal's lattice sites. The way Pauli's principle is dealt with is by comparing the one-electron band spectrum with the Fermi level of an ideal free-electron gas (see Nomenclature). The Fermi level is the chemical potential of such a gas. The exclusion principle can make it so high that for white dwarfs and neutron stars, the pressure it generates prevents the system from becoming a black hole. But the quantum correlation we are concerned with in this section is not Pauli's principle but the overlap between atomic orbitals, usually neglected in simple tight-binding calculations of band structure. The main assumption of the *tight-binding approach* to band spectra is that atoms in a crystal interact only very weakly. As a consequence, the electron's state vector should not differ very much from that of the plain juxtaposition of atomic orbitals located at the crystal's lattice sites. However, neglecting almost all interaction terms and overlap integrals (atomic states at different lattice sites need not be orthogonal to each other) may be too drastic an approximation. Thus Section 2 is devoted to a thorough discussion of the issue.

Instead, the framework of Section 3 is that of *second quantization*. Again, our view of the crystal is that of tight-binding (atoms do not lose their identities).

---

[2] Sadly, the generalized disbelief in the mere existence of atoms just one century ago may have contributed to Ludwig Boltzmann's suicide.

Here we are indeed concerned with Pauli's principle. But we deal with it in the style of quantum field theory, by allowing *at most one* electron of each spin projection per atom. For an electron to move ("hop") one lattice site, it must be *annihilated* at its former host atom and *created* in its nearest neighbor one. The purpose of this section is to illustrate an efficient Monte Carlo scheme that implements this strategy to find the ground state of many-electron systems. Recognizing that electrostatic (Coulomb) interaction between electrons is *not* a weak effect but is simply overwhelmed by Pauli's principle, a popular model of itinerant magnetism (the Hubbard model) adds to its Hamiltonian a repulsion term whenever an atom hosts *two* (opposite spin projection) electrons.

Section 4 explores the boundaries of the concept of solid. Perhaps, it should be regarded as a metaphor of this concept. We illustrate a non-equilibrium spatiotemporal pattern formation process, akin to resonant crystal structures, in arrays of FitzHugh-Nagumo cells.

2. Band spectra in the tight-binding approach: effects of the overlaps between neighboring orbitals

2.1 Quantum mechanics in a nutshell

For the benefit of those readers who are unfamiliar with the standard formalism of quantum mechanics, we review its main facts:

- **Dynamical states are vectors**: one can account for the wavelike behavior of quantum objects (e.g., diffraction of single electrons by two slits) by letting their dynamical state $|\psi\rangle$ belong to a vector space over the complex numbers. In few problems (e.g., addition of angular momenta), this vector space is finite-dimensional. But most problems entail infinite sequences (e.g., energy spectrum of the hydrogen atom) or even a continuum of values (e.g., in the measurement of positions and momenta), so the notion of dimension is replaced by that of *completitude* (*any* state can be spanned in suitable "bases"). By assigning a complex number $\langle\varphi|\psi\rangle$ (their "internal product") to every pair of dynamical states $|\psi\rangle$, $|\varphi\rangle$, the complete vector space is made into a *Hilbert* space.

- **Probabilistic interpretation**: if $|\psi\rangle = \sum_I \alpha_I |\psi_I\rangle$ (be aware that the index set I may be infinite or may even be a patch of $\mathbb{R}^d$), then $|\alpha_I|^2$ yields the probability to find an outcome represented by $|\psi_I\rangle$, when the system is in state $|\psi\rangle$. This obviously requires normalization: $\langle\psi|\psi\rangle = 1$.

- **Dynamical magnitudes are linear operators** L, which take a vector into another vector. For instance, the projector $P_\varphi := |\varphi\rangle\langle\psi|$ projects state $|\psi\rangle$ onto $|\varphi\rangle$. Measuring a dynamical magnitude thus means finding one of its eigenvalues $L|l_I\rangle = l_I|l_I\rangle$. Also of interest is the mean (or expectation) value $\langle\psi|L|\psi\rangle$ of L in a generic state $|\psi\rangle$. Correspondence with classical physics imposes that those eigenvalues be real, and thus dynamical magnitudes must be self-adjoint (Hermitian) operators (P_φ is thus *not* a dynamical magnitude).

- **Unitary evolution**: in order to conserve the probabilistic interpretation, the dynamic evolution of the state is accomplished by a unitary operator. Again, correspondence with classical physics (already implicit in Schrödinger's equation) forces this operator to be $\exp(-iH)$.

- **Wave function**: a possible "basis" set is that of eigenstates ($X|x\rangle = x|x\rangle$) of the position operator, namely, $|\psi\rangle = \int dx\, \psi(x)|x\rangle$. The *wave function* $\psi(x)$ plays here the role of the coefficients α_I. In modern notation, $\psi(x)$ is written as $\langle x|\varphi\rangle$, so one writes $|\varphi\rangle = \int dx\, |x\rangle\langle x|\varphi\rangle$.

- **Orthogonality** ($\langle\varphi|\psi\rangle = 0$): a given eigenvalue l_1 of a Hermitian operator L may have a single eigenstate $| l_1\rangle$ (the normalized one out of a dimension-1 subspace over the complex numbers) or more (in this case, it is said to be *degenerate*). Eigenstates corresponding to *different* eigenvalues are *automatically* orthogonal.

2.2 Naive tight-binding approach to band theory

As argued in Section 1, the starting point of this approach is to express the electron's state vector as a linear combination of atomic orbitals (LCAO) located at the crystal's lattice sites (we illustrate the procedure in 1D, but clearly, it can be extended to any dimension and lattice symmetry). The eigenvalue problem of the isolated atom centered at x_c is $H_{\text{atom}}\, |\psi_{\text{atom}}\rangle = E_{\text{atom}}\, |\psi_{\text{atom}}\rangle$, with $H_{\text{atom}} = T + V(x - x_c)$. We then place a copy $|i\rangle$ of $|\psi_{atom}\rangle$ centered at each lattice site i ($x_i = ia$) and write the electron's state in the crystal as LCAO

$$|\psi_{\text{crystal}}\rangle = \sum_i c_i|i\rangle \quad (1)$$

(clearly, $c_i = \langle i|\psi_{\text{crystal}}\rangle$). Now, even though the interatomic distance in the crystal (the "lattice spacing" a) is usually larger than the range x_0 of the atomic orbitals, the atomic cores *do* interact, and one should include at least two effects:

- A *correction* to the isolated atomic level E_{atom} (we shall call α the corrected level)

- Electron tunneling between neighboring orbitals (let γ be a gauge of the energy involved in such a "hopping" process)

It thus makes sense to write up the lattice Hamiltonian in terms of projection operators as

$$H_{\text{crystal}} = \alpha \sum_i |i\rangle\langle i| - \gamma \sum_i (|i+1\rangle\langle i| + |i\rangle\langle i+1|). \quad (2)$$

Two timely comments are:

1. The presence of $|i\rangle\langle i+1|$, the adjoint of $|i+1\rangle\langle i|$, ensures that H_{crystal} be *Hermitian*.

2. The minus sign in the second term ensures crystal stability (energy is released by forming a crystal).

Using Eqs. (1) and (2), the eigenvalue problem $H_{\text{crystal}}\, |\psi_{\text{crystal}}\rangle = E_{\text{crystal}}\, |\psi_{\text{crystal}}\rangle$ for the electron in the crystal reads

$$[\alpha \sum_i |i\rangle\langle i| - \gamma \sum_i (|i+1\rangle\langle i| + |i\rangle\langle i+1|)] \sum_j c_j|j\rangle = E_{\text{crystal}} \sum_j c_j|j\rangle. \quad (3)$$

Assuming the states $|i\rangle$ to be *orthogonal* to each other, the left-hand side of Eq. (3) reads $\sum_i c_i|i\rangle - \gamma\sum_i (c_i|i+1\rangle + c_{i+1}|i\rangle)$. If the number of sites in the crystal is large enough (usually it is $\sim 10^6$), one can greatly simplify the problem by assuming periodic boundary conditions (PBC). This allows to rearrange the sums (their indices become dummy), and Eq. (3) reads $\sum_i \left[(E_{\text{crystal}} - \alpha)c_i - \gamma\sum_i (c_{i-1} + c_{i+1})\right]|i\rangle = 0$. Clearly, the LCAO assumes that the $|i\rangle$ are *linearly independent* (be they orthogonal or not), so we are left with the system of difference equations:

$$(E_{crystal} - \alpha)c_i - \gamma\sum_i (c_{i-1} + c_{i+1}) = 0, i = 1 \dots N \equiv 0. \tag{4}$$

Again invoking PBC, one tries the form $c_j = \exp ijka$ with $-\pi < ka \leq \pi$ (Bloch phase factors) and obtains the known cosine spectrum

$$E_{\text{crystal}} = \alpha - 2\gamma\cos ka, \quad -\pi < ka \leq \pi. \tag{5}$$

What has been left behind? Much indeed:

- We know that α equals E_{atom} plus some correction, but we do not know what the correction is.
- Similarly, we know that γ is the expectation value of the effective potential $W_i := \sum_{j\neq i} V(x - x_j)$ felt by an electron at $x \sim ia$, due to the presence of other atoms. We have kept just $j = i \pm 1$, but even in this approximation, we do not know what the correction is.
- To what extent can one assume the states $|i\rangle$ to be *orthogonal* to each other? This assumption is correct in the absence of interatomic interaction, but not necessarily when atoms interact.

2.3 Tight-binding band calculation: properly done

Recognizing that $H_{\text{crystal}} = \sum_i (H_i^{\text{atom}} + W_i)$ and using Eq. (1), E_{crystal} turns out to be [1, 2]

$$E_{\text{crystal}} = E_{\text{atom}} + \left[\sum_i \alpha_i|c_i|^2 + \sum_{ij}\gamma_{ij}c_i^* c_j\right] \Big/ \left[\sum_i |c_i|^2 + \sum_{ij} S_{ij}c_i^* c_j\right], \tag{6}$$

where

$$\alpha_i := H_{ii} = \langle i|W_i|i\rangle, \gamma_{ij} := H_{ij} = \langle i|W_i|j\rangle, S_{ij} := \langle i|j\rangle, j \neq i. \tag{7}$$

The contribution of the S_{ij} (known as *overlap integrals*) to the band spectrum is our main concern in this section. But not less interesting are that of the α_i terms—which, as argued, shift the electronic energy in an atom from its isolated value E_{atom}, as a collective effect of the other atoms—and that of the γ_{ij}. The latter can be regarded as the sum of two contributions, as $V_j := V(x - x_j)$ can be singled out from W_i. Then whereas the two-center integrals $\gamma_{ij}^{(2)} := \langle i|V_j|j\rangle$ involve only sites i and j, the three-center integrals $\gamma_{ij}^{(3)}$ also involve the sum $\sum_{l\neq i,j} V(x - x_l)$ of the potentials of the remaining atoms in the solid. Hence, the $\gamma_{ij}^{(3)}$ can be interpreted as the collective effect on the overlap between orbitals i and j.

Variation of Eq. (6) with respect to the LCAO coefficients of Eq. (1)—namely, $\partial E_{\text{crystal}}/\partial a_j^* = 0$—yields $(H_{ij} - E_{\text{crystal}} S_{ij}) a_j = 0, \forall j$. Assuming PBC, H_{ij} and S_{ij} are functions only of the interatomic distance na, with $n = |i - j|$. Again using $a_n = \exp inka$ with $-\pi < ka \leq \pi$, Eq. (6) yields

$$E_{\text{crystal}} = E_{\text{atom}} + \left[\alpha + 2\sum_n H_n \cos nka\right] \Big/ \left[1 + 2\sum_n S_n \cos nka\right]. \tag{8}$$

Note however that the number of multicenter integrals to be computed is immense! Because of that, most tight-binding calculations plainly ignore almost all the multicenter integrals (keeping only those involving nearest neighbors) and neglect orbital non-orthogonality. This way, the familiar cosine spectrum is obtained. Often, multicenter integrals are just regarded as parameters to fit the results of more sophisticated calculations made by other methods at the highest symmetry points of the Brillouin zone.

In the following, we compute *all* the multicenter integrals *exactly* in the framework of a simple model for the atomic potential. The results help get an intuition on the effect on band spectrum of neglecting overlap integrals and distant-neighbor interactions.

2.4 A simple model that yields an exact tight-binding band spectrum

We restrict ourselves to a 1D monoatomic crystal and assume the interatomic distance a to be larger than the effective range of the screened Coulomb potential representing the atomic core. In such a situation, we can approximate the latter by a Dirac δ-function (complete screening up to the scale of the nucleus):

$$V_{\text{crystal}}(x) = -V_0 \textstyle\sum_n \delta(x - na). \tag{9}$$

The solution to $H_{\text{atom}} |\psi_{\text{atom}}\rangle = E_{\text{atom}} |\psi_{\text{atom}}\rangle$, with $H_{\text{atom}} = -\frac{\hbar^2}{2m}\frac{d^2}{dx^2} - V_0\delta(x)$, is an exponential function of the form $\psi_{\text{atom}}(x) = \langle x|\psi_{\text{atom}}\rangle = x_0^{-\frac{1}{2}} \exp\left(-\frac{|x|}{x_0}\right)$. Its range x_0 is related to E_{atom} by$-E_{\text{atom}} = \hbar^2/2mx_0^2 = mV_0/2\hbar^2$.

The only two spatial scales involved in this problem are x_0 and the lattice spacing a. The parameter $t = a/x_0$ will thus allow us to follow the formation of energy bands (k-space picture) as atoms get close together (real-space picture). All the multicenter integrals can be computed analytically in terms of t. The results are $S_n = (1 + nt)\exp(-t)$, $\alpha = 2E_{\text{atom}} \exp(-t)/\sinh t$, and $\gamma_n = 2E_{\text{atom}}[n + \exp(-t)/\sinh t] \exp(-nt)$ [3]. We thus get the following closed expression for $\lambda := (E_{\text{crystal}} - E_{\text{atom}})/E_{\text{atom}}$:

$$\lambda(k,t) = [A_0(t) + A_1(t)\cos ka]/[1 + S(t)\cos ka], \quad -\pi < ka \leq \pi, \tag{10}$$

where $A_0 = \exp(-t)\sinh t/[\sinh t \cosh t - t]$, $A_1 = \sinh t/[\sinh t \cosh t - t]$, and $S = [t\cosh t - \sinh t]/[\sinh t \cosh t - t]$ [3].

Explicit evaluation of Eq. (10) at the bottom ($ka = 0$) and top ($ka = \pi$) of the band shows that for $t < 4$, the cosine spectrum of Eq. (5) underestimates both. Moreover, the multicenter integrals neglected in the cosine spectrum shift *unevenly* the top and bottom of the exact spectrum. Hence, the approximation performs worse for the top than for the bottom of the band.

3. Quantum Monte Carlo method for systems with strongly correlated fermions

3.1 Quantum *statistical* mechanics in a nutshell

The *state vectors* dealt with in Section 1 represent *pure* states. They are the ones which display the spectacular effects seen in recent experiments. Since in this section, we will allow *creation annihilation* of electron states, we must work in the framework of the *grand canonical ensemble*.[3] When one deals with *statistical ensembles* of quantum states, the object of interest is the Hermitian operator $\exp(-\beta H)$, called the *density matrix* operator (here $\beta := (k_B T)^{-1}$ and $k_B = R/N_A$ are Boltzmann's constant).

What drives our interest in the *density matrix*—namely, the matrix elements between pure states of $\exp(-\beta H)$—is the fact that it can be used to find the ground state of many-body systems by stochastic methods. For β large enough, $\exp(-\beta H)$ acts effectively as a *projector* over the lowest-lying energy eigenstate to which the initial (trial) state $|\varphi\rangle$ is not definitely orthogonal. Let E be the corresponding eigenvalue, and consider another trial state $|\chi\rangle$ over which we will project the result. Then we may numerically compute E from

$$\exp(-\Delta\beta E) = \lim_{\beta\to\infty} [\langle\chi|\exp[-(\beta+\Delta\beta)H]|\varphi\rangle / \langle\chi| \exp(-\Delta\beta H)|\varphi\rangle]. \tag{11}$$

But what is yet more interesting is that in the process, we find a good estimate of the eigenstate itself, namely, its composition in terms of a known basis.

3.2 Monte Carlo pursuit of the ground state

The first step in this computation is to divide the interval $[0, \beta]$ into L "time" slices of width $\Delta\tau = \beta/L$. Some comments are in order:

a. We take our language from the formal analogy between the density matrix and the evolution operators.

b. Note that in our case, $\exp(-\beta H)$ is *not meant to be traced over* as it should be in a thermodynamic calculation: here it must rather be considered as a formal tool to make sense in the limit $\beta \to \infty$.

c. We may call $U = \exp(-\Delta\tau H)$ the *transfer matrix* operator.

If we can decompose H into a sum of several terms H_i which (although not commuting among them) are themselves *sums of commuting terms*, then for L large enough, the error of approximating

$$U = \exp[-\Delta\tau(H_1+2)] = \exp(-\Delta\tau 1)\exp(-\Delta\tau H_2)\exp\left\{-(\Delta\tau)^2[H_1,H_2]\right\}$$
$$= U_1U_2\left\{1-(\Delta\tau)^2[H_1,H_2]+\ldots\right\} \sim U_1U_2$$

would be at most of order $(\Delta\tau)^2$. Hence

[3] We have already stated that the Fermi level is the *chemical potential* of an ideal free-electron gas. This concept is peculiar of the *grand canonical ensemble*.

$$\langle\chi|\exp(-\beta H)|\varphi\rangle \sim \left\langle\chi|(U_1U_2)^L|\varphi\right\rangle. \tag{12}$$

In order to evaluate expression (12), we introduce complete sets of states at each time slice.

The clue to quantum Monte Carlo simulation of Eq. (11) resides in evaluating the sums over complete states *by importance sampling*: in order to do that, observe first that we can rather arbitrarily decompose

$$\left\langle\psi_j|(U_1U_2)^L|\psi_i\right\rangle = S_{ij}P_{ij} \tag{13}$$

as the product of a probability times a (complex) number which we will call a "score." The probability distribution P_{ij} is at our disposal in order to optimize numerical convergence, minimize statistical error, etc. It can be shown that the way to achieve the last goal is by assigning to every matrix element the same score: that is the basis for the so-called *population method*. Here the initial (trial) state is represented by a "population" in which there are n_i copies of state $|\psi_i\rangle$. The latter corresponds to a definite assignment of occupation numbers both in coordinate and spin (always belonging to the Hilbert space of the problem, i.e., compatible with the conserved quantum numbers). To each individual in the population, we apply the evolution operator, thus obtaining a new state after one time slice. That particular matrix element can be decomposed as indicated in Eq. (20) (but being now $S_{ij} = S = \text{const}$). The way in which we implement the P_{ij} is by making as many copies of that particular resulting state as indicated by $\langle\psi_j|(U_1U_2)^L|\psi_i\rangle/S$. Proceeding this way, we will get a different population after each time slice which we expect to approach successively to one representing the lowest reachable energy eigenstate.

We have not said anything about the way in which we evaluate the alluded matrix elements, besides the fact that we resort to the decomposition (20): if, as we have already assumed, the term H_i can itself be decomposed into mutually commuting terms, we need only to focus on the Hilbert space of that (much smaller) system. We can compute exactly the matrix elements of the evolution operator for that cluster, write them as the product of a probability times a score (now we can choose the probability distribution to minimize total computing time), and make transitions among cluster states according to those probabilities, assigning then the corresponding score to the particular transition.

3.3 The case of fermions

Again within the *tight-binding approach* to crystalline solids, quantum *creation* ($c^{\dagger}_{is}$) and *annihilation* (c_{is}) operators determine the existence of electrons with spin projection σ at site i. For the state vector of the whole set of electrons in the crystal to be totally antisymmetric under exchange, those operators must *anticommute* with each other, unless they refer to the same site and spin projection. In such a case, there can be *at most one* electron per site and spin projection, as required by Pauli's principle.

In the case of the 1D Hubbard model, we chose the following decomposition of the Hamiltonian, which allows us to consider clusters of only two sites:

$$H_1 = -t\textstyle\sum_{oddj}\sum_\sigma\left(c^{\dagger}_{j+1\sigma}c_{j\sigma} + h.c.\right) = -t\displaystyle\sum_{oddj}\sum_\sigma h_{j,j+1}$$

$$\mathrm{H}_2 = -t\textstyle\sum_{\mathrm{evenj}}\sum_\sigma\left(\mathrm{c}^{\dagger}_{\mathrm{j+1}\sigma}\mathrm{c}_{\mathrm{j}\sigma} + \mathrm{h.c.}\right) = -\mathrm{t}\displaystyle\sum_{\mathrm{evenj}}\sum_\sigma \mathrm{h}_{\mathrm{j,j+1}} \tag{14}$$

$$H_3 = -t\sum_{\text{all } j} n_{j\uparrow} n_{j\downarrow}.$$

The corresponding matrix elements are then $\langle \psi_{i+1} | U_3 U_2 U_1 | \psi_i \rangle$ with $U_1 = \prod_{oddj} \exp\left(-\Delta\tau h_{j,j+1}\right)$, $U_2 = \prod_{evenj} \exp\left(-\Delta\tau h_{j,j+1}\right)$, $U_3 = \prod_j \exp\left(-\Delta\tau\, n_{j\uparrow} n_{j\downarrow}\right)$, and

$$\begin{aligned} \langle 01| \exp\left(-\Delta\tau h_{j,j+1}\right)|01\rangle &= \langle 10| \exp\left(-\Delta\tau h_{j,j+1}\right)|10\rangle = \cosh \Delta\tau, \\ \langle 10| \exp\left(-\Delta\tau h_{j,j+1}\right)|01\rangle &= \langle 01| \exp\left(-\Delta\tau h_{j,j+1}\right)|10\rangle = \sinh \Delta\tau, \\ \langle 00| \exp\left(-\Delta\tau h_{j,j+1}\right)|00\rangle &= \langle 11| \exp\left(-\Delta\tau h_{j,j+1}\right)|11\rangle = 1, \end{aligned} \tag{15}$$

from which we write up the (a priori) transition probabilities. Then, in case there is only one occupied site in the block, we draw a random number r and compare it with the a priori transition probability p for the state to remain the same. In case that $r > p$, we make a hopping, i.e., exchange empty and occupied states in the block.

The a priori probabilities can be better chosen if we take into account the occupation of those same two sites by electrons with the other spin projection, thus anticipating to the fact that they will penalize doubly occupied sites [4, 5]. This will certainly improve convergence.

4. Non-equilibrium routes to soft solids

Up to now, we have dealt with crystalline solids. This means that disregarding the *topology*[4] of the interaction network, we paid attention to the underlying *geometry* of the quantum problem. At present, a host of synthetic materials has outperformed metals at their initial tasks. Some of them still display a varying degree of crystalline character, but others are not crystalline at all. Vulcanized rubbers are an example: created by forcing random chemical bonds in a melt (a "spaghetti dish"), they are inhibited to flow, and, thus, they are amorphous solids.[5] But they exhibit a varying degree of viscoelastic behavior. In the last decades, the vast discipline of *soft condensed matter* has incorporated to mainstream research in solid-state physics, at equal footing with crystalline solids. The scope of soft condensed matter is very wide. In particular, it considers many non-equilibrium routes to *self-assembled emergent* structures. Of huge interest is the neocortex (not just because understanding the brain's behavior is one of the "Holy Grails" of science, but because in doing it we may achieve to master a computational strategy which is far more efficient than the present one).

We devote this section to the emergence of non-equilibrium routes to spatio-temporal patterns in an assembly of model "neurons" which keep their essential trait, namely, *excitability*. Admittedly, here the interaction network has the *topology* of a lattice, but here it is not the underlying geometry that is at stake. What does matter here is that the boundary condition be compatible with the interaction, a fact that contributes to the network's topology.

---

[4] It will be a lattice only if all interactions but nearest neighbor ones are neglected. Note that crystals may even have a Cayley tree structure, like the so-called "Bethe lattices."

[5] The electronic properties of amorphous solids are also of interest, e.g., in the photovoltaic (PV) industry.

4.1 The non-equilibrium potential (NEP)

It is often hard to tell to what extent an innovation embodies a paradigm shift, for the high diversity (both in scope and extent) of innovations. The formalism of quantum mechanics can be regarded as such—with respect to the Newtonian paradigm—despite the strict correspondence between commutator and Poisson bracket Lie algebras. Also can Einstein's three papers in his "annus miraculus" be considered as such, for they demolished our former conceptions of time, of the nature of particles and waves, and of a clockwork universe. In 1908, Paul Langevin supplemented the Newtonian paradigm by letting the forces be of stochastic nature [6]. It is up to your taste to call this innovation a paradigm shift: it definitely abolished our clockwork universe conception and opened up a new chapter in the theory of differential equations. The resulting paradigm is well suited to the current situation, urged by the challenges of nanoscience (where the "systems" are submitted to strong ambient fluctuations) and favored by the increasing parallelism of computational architectures (the simulation schemes are essentially local).

The modern approach to continuous-time dynamic flows is of *first order*.[6] Given an initial state x_i of a continuous-time, dissipative, autonomous dynamic flow $\dot{x} = f(x)$, its conditional probability density function (PDF) $P(x,t|x_i,0)$ when submitted to a (Gaussian, centered) white noise $\xi(t)$ with variance γ, namely,

$$\dot{x} = f(x) + \xi(t), \text{ with } \langle\xi(t)\rangle = 0 \text{ and } \langle\xi(t)\xi(t')\rangle = 2\gamma\delta(t-t') \tag{16}$$

obeys the Fokker-Planck equation (FPE):

$$\partial P(x,t|x,0) + \partial J(x,t|x,0) = 0, \text{ with } J(x,t|x,0) = D^{(1)}P - \partial_x\left[D^{(2)}(x)P\right] \tag{17}$$

in terms of the "drift" $D^{(1)} = f(x)$ and "diffusion" $D^{(2)} = \gamma$ Kramers-Moyal coefficients. Being the flow nonautonomous but dissipative, one can expect generically situations of statistical energy balance in which the PDF becomes stationary, $\partial_t P_{st}(x) = 0$, thus *independent of the initial state*. Then by defining the *non-equilibrium potential* $\Phi(x) := -\int_{x_0}^{x} f(y)dy$, it is immediate to find

$$P_{st}(x) = N(x_0)\exp\left[-\Phi(x)/\gamma\right]. \tag{18}$$

For n-component dynamic flows, $\Phi(\mathbf{x})$ is defined as $-\lim_{\gamma\to 0}\gamma\ln P_{st}(\mathbf{x};\gamma)$ [7], but finding it ceases to be a straightforward matter.[7] The purpose of this section is to illustrate its *usefulness* when known. It is a *Lyapunov function* for the deterministic dynamics, and the barriers for activated processes can be straightforwardly computed $\lim_{\gamma\to 0}\gamma\ln$.

4.2 The FitzHugh-Nagumo model and its NEP

Neurons communicate with each other through "action potentials," which are pulsed variations in the polarization of their membranes. The celebrated Hodgkin-Huxley model of neural physiology was one of the great scientific achievements of the past century. When the goal is insight, however, it is too cumbersome a model

---

[6] Recall it was Hamilton who first succeeded in casting conservative systems as first-order ones. In so doing, he put *coordinates* and *momenta* on the same footing. Systems are *conservative* if their phase space does not contract.

[7] A key is to ensure the multidimensional version of $D^{(2)}$ (a symmetric tensor) to be *nonsingular*.

to work with. A caricature of this model which nonetheless stresses its essence is thus far more desirable in many situations. The FitzHugh-Nagumo model is the minimal model capable to produce *action potentials*, and the key to this behavior is *excitability*. In its minimal expression, the FHN model reads

$$\begin{aligned} \dot{u} &= f(u) - v, \\ \dot{v} &= \epsilon(\beta u - v). \end{aligned} \tag{19}$$

The *activator* field u relaxes very fast and displays *autocatalytic* dynamics (the more there is, the more it produces, but in a *nonlinear* fashion) as needed to produce an action potential. Its *nullcline* $v = f(u)$ (the locus of $\dot{u} = 0$) is a decreasing S-shaped (typically cubic) curve. On the other hand, the *inhibitor* or recovery field v relaxes very slowly (it mimics the time-dependent conductance of the K^+ channels in the axon membrane), so in the end, it enslaves the dynamics. Parameter ϵ is usually very large, to account for the large difference in relaxation rates. Calling λ_1 and λ_2 the eigenvalues of the diffusion tensor, the NEP for the *autonomous* system described by Eq. (19) is [8]

$$\Phi(u, v) = {\lambda_2}^{-1}(\beta u - v)^2 + (\lambda_1 \epsilon)^{-1}\left(\beta u^2 - 2\int_{u_0}^{u} f(x)dx\right)^2. \tag{20}$$

For *nonautonomous* cases, one can draw consequences from Eq. (20) as far as the driving is much slower than the involved relaxation times (adiabatic approximation). In the following, we exploit this advantage.

4.3 Arrays of excitable elements

The result (20) has been employed [9–13] to find the optimal noise variance γ for arrays of excitable elements to display *stochastic resonance synchronized behavior* (see Nomenclature). Here, we briefly illustrate one such a case, where the coupling is *inhibitory* (when neuron i fires, neurons $i \pm 1$ are less likely to fire) [14]. Inhibitory coupling is central in the dynamics of neocortical pyramidal neurons and cortical networks, and plays a major role in synchronous neural firing. On the other hand, inhibitory interneurons are more prone to couple through *gap junctions* (diffusive or "electric" coupling) than excitatory ones. In the transition from wake to anesthetic coma, for instance, diffusive coupling of inhibitor fields helps explaining the spontaneous emergence of low-frequency oscillations with spatially and temporally chaotic dynamics.

We consider a ring of N identical excitable FHN cells, with their *inhibitor* fields *electrically* coupled to those of their nearest neighbors. The system is moreover submitted to a common *subthreshold* (see Nomenclature) harmonic signal $S(t)$ and *independent* additive Gaussian white noises in each component and each site, all with the same variance γ.

Numerical simulation of this stochastic system with increasing γ—for appropriate values of the diffusive coupling E between neighboring inhibitor fields—reveals the noise-induced phenomena taking place: *synchronization with the external signal* of the ring's activity and (imperfect) *spatiotemporal self-organization* of the cells. For an optimal value of γ, a *stochastic resonance* phenomenon takes place, and the degree of *spatiotemporal self-organization*—alternancy between two *antiphase states* (APS)—is maximum.

For very low γ, only small-amplitude and highly homogeneous $[u_i(t) \approx u_j(t)]$ subthreshold oscillations (induced by the adiabatic signal) occur around the $S = 0$

rest state. As γ increases, so does the number of cells that become noise-activated during roughly half a cycle of the external signal. For γ even higher, the cells' activity enhances its coherence with the external signal as a consequence of its coupling-mediated self-organization: as one neuron activates, it usually inhibits its nearest neighbors. The outcome of this phenomenon is the APS, which partially arises along the ring during the stage of activation by noise. In this scenario, noise (together with coupling and signal) plays a constructive role. Nonetheless for γ too large, the sync becomes eventually degraded.

4.4 Spatiotemporal pattern formation in arrays of FHN neurons

We exploit the knowledge of the NEP in Eq. (20) to attempt an analytical description of the problem in Section 4.3. The case of *perfect* spatiotemporal self-organization would be equivalent to a two-neuron system with variables u_1, u_2, v_1, and v_2 and PBC. This simple model allows the formation of an *antiphase* state. Since a NEP cannot be easily found for this system—and with the only purpose of calculating barrier heights—we further reduce this description by *projecting the dynamics along the corresponding slow manifolds*:

$$\epsilon\beta u_{1,2} - v_{1,2} + 2E\,(v2 + v1 - v_{1,2}) = 0. \tag{21}$$

The projected two-variable system turns out to be gradient, a situation in which a NEP can always be found. As a consequence of the PBC, the NEP landscape along the slow manifolds is symmetric with respect to the $u_1 = u_2$ line. For $E = 0.5$ and maximum signal amplitude, the system has two *uniform* attractors (both cells inhibited, both cells activated), two APS (with one cell activated and one inhibited) with the *same* value of $\Phi(u_1, u_2)$, four saddles, and one maximum. For $S = 0$ instead, the uniform attractor with both cells activated has collapsed with the maximum, and, hence, two saddles have disappeared.

When the value of $\Phi(u_1, u_2)$ at the uniform attractor, either APS and either corresponding saddle, is plotted as a function of S, one can see the following:

- Near maximum signal, the uniform attractor yields its stability to the APS. From this value of S on, the NEP barrier for the uniform attractor to decay into the APS (a noise-activated process) is small enough.

- Way before minimum signal, each APS collapses with its own saddle.

One then understands the picture: as S increases, whatever of the APS is chosen. As S decreases past the collapse, only the uniform attractor survives. However, the neuron which was activated before has not recovered completely. Hence in the next signal cycle, the other APS is more likely to appear.

5. Conclusions

In Sections 2 and 3, we have discussed the influence of quantum correlations on the formation of tightly bound solids. Section 2 is devoted to the effects of the overlaps and neglected multicenter integrals on tight-binding band spectra. An exact calculation in the framework of a simple atomic model has shown that they shift *unevenly* the top and bottom of the band spectrum (their effects are more pronounced at the top). Section 3 introduced a quantum Monte Carlo method specific for strongly correlated fermion systems. Section 4 addressed the stochastic dynamics of a ring of FHN cells—with nearest neighbor *electric* (diffusive) coupling between their *inhibitor* fields—undergoing spatiotemporal pattern formation induced by noise and coupling. By means of a simple model for which a NEP can be found, the mechanism whereby the process takes place was investigated analytically.

Acknowledgements

The author is deeply indebted with his coauthors G.G. Izús, A.D. Sánchez, and M.G. dell'Erba from IFIMAR-CONICET (Faculty of Exact and Natural Sciences) and D.A. Mirabella and C.M. Aldao from INTEMA-CONICET (Faculty of Engineering) of the National University of Mar del Plata (UNMdP), Argentina, with whom he undertook part of the work referred to here. Support by UNMdP, through Grant EXA826–15/E779, is acknowledged.

Nomenclature

Ideal gas	the (identical) particles composing such a gas do not interact between themselves.
Free electrons	they are not submitted to any external (e.g., crystal) potential.
Chemical potential	it is the cost of adding a particle to the system. For two open systems (which can exchange matter and energy with their environments) to come to equilibrium, not only their temperatures but their chemical potentials must be equal.
Subthreshold	unable by itself to drive a transition.
Stochastic resonance	nonlinear systems may display the property of amplifying a *subthreshold* input signal in the presence of noise with the right intensity.

Author details

Roberto Raúl Deza
IFIMAR, Faculty of Exact and Natural Sciences, National University of Mar del Plata, CONICET, Argentina

*Address all correspondence to: deza@mdp.edu.ar

References

[1] Mirabella DA, Aldao CM, Deza RR. Orbital nonorthogonality effects in band structure calculations within the tight-binding scheme. American Journal of Physics. 1994;**62**:162-166. DOI: 10.1119/1.17637

[2] Mirabella DA, Aldao CM, Deza RR. Effects of orbital nonorthogonality on band structure within the tight-binding scheme. Physical Review B: Condensed Matter. 1994;**50**:12152-12155. DOI: 10.1103/PhysRevB.50.12152-12155

[3] Mirabella DA, Aldao CM, Deza RR. Exact one-band model calculation using the tight-binding method. International Journal of Quantum Chemistry. 1998; **68**:285-291. DOI: 10.1002/(SICI)1097-461X(1998)68:4<285::AID-QUA6>3.0.CO;2-R

[4] Kung D, Dahl D, Blankenbecler R, Deza RR, Fulco JR. New stochastic treatment of fermions with application to a double-chain polymer. Physical Review B. 1985;**32**:2022-2029. DOI: 10.1103/PhysRevB.32.2022

[5] Braunstein LA, Deza RR, Mijovilovich A. Exact versus quantum Monte Carlo analysis of the groundstate of the one-dimensional Hubbard model for finite lattices. In: Cordero P, Nachtergaele B, editors. Nonlinear Phenomena in Fluids, Solids and Other Complex Systems. Amsterdam: North-Holland; 1991. pp. 313-327. DOI: 10.1016/B978-0-444-88791-7.50024-7

[6] Lemons D. Paul Langevin's 1908 paper "On the theory of Brownian motion" ("Sur la théorie du mouvement brownien," C. R. Acad. Sci. (Paris) 146, 530-533 (1908)). In: AIP Conference Proceedings, Vol. 65. 1997. pp. 1079-1081. DOI: 10.1119/1.18725

[7] Graham R. Weak noise limit and nonequilibrium potentials of dissipative dynamical systems. In: Tirapegui E, Villarroel D, editors. Instabilities and Nonequilibrium Structures. Dordrecht: D. Reidel; 1987. pp. 271-290. DOI: 10.1007/978-94-009-3783-3_12

[8] Izús GG, Deza RR, Wio HS. Exact nonequilibrium potential for the FitzHugh–Nagumo model in the excitable and bistable regimes. Physical Review E. 1998;**58**:93-98. DOI: 10.1103/PhysRevE.58.93

[9] Izús GG, Deza RR, Wio HS. Critical slowing-down in the FitzHugh–Nagumo model: A non-equilibrium potential approach. Computer Physics Communications. 1999;**121-122**: 406-407. DOI: 10.1016/S0010-4655(99)00368-9

[10] Wio HS, Deza RR. Aspects of stochastic resonance in reaction–diffusion systems: The nonequilibrium-potential approach. European Physical Journal: Special Topics. 2007;**146**:111. DOI: 10.1140/epjst/e2007-00173-0

[11] Izús GG, Deza RR, Sánchez AD. Highly synchronized noise-driven oscillatory behavior of a FitzHugh—Nagumo ring with phase-repulsive coupling. AIP Conference Proceedings. 2007;**887**:89-95. DOI: 10.1063/1.2709590

[12] Izús GG, Sánchez AD, Deza RR. Noise-driven synchronization of a FitzHugh–Nagumo ring with phase-repulsive coupling: A perspective from the system's nonequilibrium potential. Physica A: Statistical Mechanics and its Applications. 2009;**388**:967-976. DOI: 10.1016/j.physa.2008.11.031

[13] Sánchez AD, Izús GG. Nonequilibrium potential for arbitrary-connected networks of FitzHugh–Nagumo elements. Physica A: Statistical Mechanics and its Applications. 2010; **389**:1931-1944. DOI: 10.1016/j.physa.2010.01.013

[14] Sánchez AD, Izús GG, Dell'Erba MG, Deza RR. A reduced gradient description of stochastic-resonant spatiotemporal patterns in a FitzHugh–Nagumo ring with electric inhibitory coupling. Physics Letters A. 2014;**378**: 1579-1583. DOI: 10.1016/j.physleta. 2014.03.048

Chapter 2

Lagrangian Quantum Mechanics: A Fully Relativistic Theory of Atomic Structure

Richard Oldani

Abstract

A fully relativistic formulation of quantum mechanics is derived by introducing a Lagrangian density of the fields between the excited and ground states and taking the action integral. The change in action, or photon, is a four-dimensional localization of fields that is defined symmetrically with respect to the field boundaries. Due to this photon model, we interpret the three mathematical formulations of atomic structure, matrix mechanics, wave mechanics, and path integrals, as different mathematical methods of describing the superposed physical components of an excited state: nucleus, electron, and photon. Recent experiments with slow and stopped light are shown to support this theoretical interpretation. The derivation of quantum theory with respect to fields requires new interpretations of the uncertainty principle, correspondence principle, complementarity, and force.

Keywords: quantum mechanics, relativity theory, uncertainty principle, correspondence principle, complementarity

1. Introduction

Nonrelativistic quantum mechanics is commonly expressed by using a Hamiltonian function, with the total energy of an atomic system given by the sum of the kinetic and potential energies. If the atom is in an excited state the energy resides at the location of the electron. The electron in an excited state has a potential energy that is converted into a photon as it decays, so emission is a statistical event that occurs at a particular point in time. Because photon creation occurs at a single point in time, the electron must be in two energy states simultaneously. The two states correspond to a single time, so emission is described by non-covariant means even though we know that it is Lorentz invariant. Although nonrelativistic methods are used to describe atomic structure, they are not a necessary requirement of quantum mechanical formalism. As Dirac stated in the first ever paper on quantum electrodynamics, "The theory is nonrelativistic only on account of the time being counted throughout as a c-number [classically], instead of being treated symmetrically with the space coordinates" [1]. The use of a continuous time parameter to describe microscopic phenomena could easily mask the way time functions in atomic structure. The practice should be closely questioned because time is the least understood of the variables in quantum mechanics. Furthermore there are grounds to support a different approach to quantum mechanics using Lagrangian mechanics. It is not only more fundamental than Hamiltonian mechanics, but if we begin with an action principle by taking the time integral of a Lagrangian, it is in agreement with special relativity theory.

2. Excitation

We begin an attempt to understand the role of continuous time in quantum mechanics by describing emission as a sequence of events that are relativistically correct. Consider a uniform distribution of atomic oscillators immersed in a coherent or partially coherent radiation field with an outer electron that occupies either of two allowable energy states $|1\rangle$ or $|2\rangle$. The radiation field is conceived of as classical, consisting of many oscillating wave train fields superimposed on the oscillators. The system may be described by a Lagrangian density that is a function of oscillator fields φ_r and radiation fields $\partial_\mu\varphi$. If Hamilton's principle is applied macroscopically to an arbitrary classical region of space-time Ω, we obtain the usual expression:

$$S(\Omega) = \int_{\Omega} L(\varphi_r, \partial_\mu \varphi) d\Omega \tag{1}$$

The behavior of the radiation field may be described by a Lagrangian density microscopically as well. The amplitude of the field at the local level fluctuates randomly, either reinforcing or canceling, thereby causing the electron to oscillate with respect to the nucleus. If the fields are of insufficient intensity to raise the electron from $|1\rangle$ to $|2\rangle$, the uncertainty relations do not apply and the electron oscillates without radiating. However, if they are sufficiently intense, the electron will resonate at the excitation energy emitting a photon with each amplitude of the wave. When this happens radiation is emitted with double the frequency of the incident wave (see **Figure 1**).

Emission by atomic oscillators is referred to as "nonlinear" if excitation is thought to be quantum mechanical or "spontaneous" if it is thought to be governed by the macroscopic Maxwell's equations [2]. The use of microscopic Maxwell's equations in a Lagrangian density allows emission to be described by a distribution of locally superposed fields that includes continuous, statistical, and quantum mechanical properties in a single model.

Due to the generality of the assumptions, examples of frequency doubling should be common in nature occurring at many energy levels. This is in fact the case, and it is most clearly evident when laser light is passed through a crystal [3]. It is also readily observed using incoherent light in scattering experiments as secondary radiation [4]. On the other hand, when an energy quantum is completely absorbed, an electron will be expelled due to the photoelectric effect, and no frequency change is observed. We see in these examples evidence that the Lagrangian provides a versatile model for the description of electromagnetic phenomena.

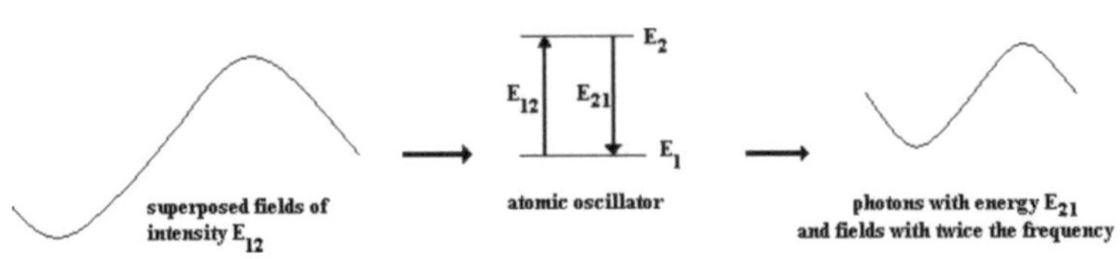

Figure 1.
Frequency doubling.

3. Decay

The first appearance of a Lagrangian in quantum mechanics is in a paper by Dirac. "We ought to consider the classical Lagrangian not as a function of the coordinates and velocities but rather as a function of the coordinates at time t and the coordinates at time t+dt" [5]. Rather than specify emission as an event that occurs at a particular point in time, Dirac is seeking compatibility with relativity theory by calculating the change in action of the electron over a space-time interval. That idea resulted in an interpretation of electron transitions as a "sum over histories" of all possible paths from one diagonalized steady state to another [6]. Although calculations made with the path integral method are accurate, they include serious theoretical problems resulting from the renormalization methods used to deal with infinities that have so far prevented a complete theory from being formulated. At a later point in the same paper, Dirac provides a possible way out of this difficulty by proposing a complementary description of quantum mechanics in terms of a "vibrating medium." The idea begins from a classical vantage point. "We introduce at each point of space-time a Lagrangian density, which must be a function of the coordinates and their first derivatives with respect to x, y, z, and t, corresponding to the Lagrangian in particle theory being a function of coordinates and velocities. The integral of the Lagrangian density over any (four-dimensional) region of space-time must then be stationary for all small variations of the coordinates inside the region, provided the coordinates on the boundary remain invariant." To obtain the quantum analogue, he divided the classical region into a number of very small subregions; however, the idea was never pursued to completion. We shall investigate that line of reasoning in more detail by describing fields over a four-dimensional region of space-time with respect to invariant field boundaries coincident with the steady states.

The excited states of an atom should be described discretely in time. As Dirac pointed out, discrete time is symmetric with the spatial coordinates allowing the emission and absorption of radiation to be described relativistically. A relativistic formulation is desirable in order to make quantum mechanics compatible with special relativity theory. To formulate a fully relativistic description of emission, we begin with a Lagrangian density $L(\phi_i, \phi_{i,\mu})$ that is a function of the coordinates and their first derivatives. Then the action integral of the Lagrangian density over a particular region of space-time must be stationary for all small variations of the continuous coordinates in the region provided the discrete coordinates on the boundary remain invariant. Let three-dimensional surfaces R_1 and R_2 representing the diagonalized coordinates be erected and then used together with the discrete time period $t_2 - t_1$ as field boundaries to define a region of space-time between $|2\rangle$ and $|1\rangle$. Integrating from $|2\rangle$ to $|1\rangle$ yields a relativistic formulation of emission that is invariant, the same for all observers:

$$S[\varphi_i(t)] = \int_{R_2}^{R_1}\int_{t_2}^{t_1} L(\varphi_i, \varphi_{i,\mu})d^3x dt = h \tag{2}$$

The end points of the electron's path are located on equipotential, space-like surfaces, and the action minimum is not equal to zero as in classical theory, but to Planck's constant h. The action $S[\varphi_i(t)]$ in (2) is functional, a function of the values of coordinates on the *discrete* boundary of the space-time surface R_2 which is in turn a function of the *continuous* space-time variables of the fields within the surface.

Solving for the action, we obtain solutions for localized energy or equivalently, photon creation; due to a transformation of field:

$$E\tau = h \tag{3}$$

Solutions of Equation (3) are determinations of energy and time between exact four-dimensional field boundaries, so they are also exact. In the case of time periods, this has been confirmed to the limits of experimental accuracy by atomic clocks that can operate for many billions of years without significant error [7]. Thus field energy from a laser is absorbed by the lattice of ytterbium atoms and localized within a four-dimensional field boundary. The emission and absorption energies E_{21} and E_{12} have also long been assumed to be exact by astronomers when employed for the measurement of distant star composition. Emission and absorption spectra, together with a red shift, often require billions of years before they are observed, thereby reflecting the precise role of energy and time in natural phenomena.

From (2), a model of atomic structure may be constructed. The three field sources present in excited atomic states, electron, proton, and photon, superpose linearly and are momentarily stabilized in steady states. Although force is an unnatural concept in quantum mechanics, it may be interpreted with respect to the field boundaries that separate point sources by a careful consideration of (2). Thus the force on a bound electron due to the potential is equal to the continuously distributed excitation energy divided by the distance between field boundaries. In order for quantum mechanical forces to agree with relativity theory, we need only require that the action integral be invariant for all potentials both free and bound.

Whereas the quantum mechanical force of bound states is due to well-defined field boundaries, the force due to instantaneous exchanges of momentum, such as occurs in the Compton effect, may be interpreted as a result of the encounter of a single, exact field boundary (x_o,y_o,z_o,t_o) with a material point. Wave properties, on the other hand, occur in free space when field boundaries have no reference point, so they cannot be observed at all. Thus we interpret complementarity as the different ways that fields and its field boundaries interact with matter.

4. Interpretation of mathematical models

4.1 Fully relativistic quantum mechanics

A fully relativistic description of excited atomic states specifies discrete four-dimensional field boundaries and continuous localized fields between the boundaries. Excited atomic states, electron, photon, and nucleus, are interpreted as a linear superposition of three field sources with respect to their electrostatic and electromagnetic field components. Recent experiments, referred to by the authors as "photon capture" and "photon storage," support the accuracy of this theoretical interpretation [8]. In these experiments light coherence is converted to atomic coherence and back again, so the photon in localized form must be present in excited atomic states from the time energy is absorbed until it is emitted. We describe the linear properties of atomic structure by introducing a wavelike field source ε, the localized photon, into our description of excited atomic states. The modified Hamiltonian is now given by

$$H = T + \varepsilon + V \tag{4}$$

and the Lagrangian is similarly given by

$$L = T + \varepsilon - V \tag{5}$$

where T represents the bound electron's energy, ε represents the energy of a "captured" photon, and V represents the potential energy due to the nucleus/proton. Each of the three field sources possesses a unique vector field, that is, a field with definite field geometry that is delimited from the others by field boundaries, where plus and minus signs indicate the linear superposition of delimited fields.

Eqs. (4) and (5) contain the essence of quantum mechanics as a three-body problem in real space as opposed to current descriptions based on a two-body system in abstract space. The use of an abstract space is necessary for nonrelativistic descriptions of atomic structure since the photon is not treated as an independent particle. The equations revert to their classical form when the influence of ε is negligible or equivalently when field boundaries are no longer determinable. To see whether the model accurately describes atomic structure, we shall compare it to the existing mathematical models.

4.2 The path integral formulation

From (5), it is postulated that the contribution of two energies is summed and one is subtracted to give the transition energy. In the path integral formulation, there are in fact two contributions that are summed, one determined by the paths and the other by the phase. In Feynman's words, "The paths contribute equally in magnitude, but the phase of their contribution is the classical action (in units of ħ)" [9]. The computation of the total energy is not complete, however, until contributions to the potential V due to self-energy are subtracted away by renormalizing. Therefore the mathematical structure of the Lagrangian in nonrelativistic quantum mechanics indicates the existence of three contributions and is in agreement with (5). It differs fundamentally from the fully relativistic Lagrangian method described here in its interpretation of space-time. Eq. (2) treats space and time equivalently as real parameters for both the integration limits and the region between them. On the other hand, the path integral formulation uses abstract forms of space and time to describe the region between the steady states since the paths follow all trajectories and for all times.

4.3 Matrix mechanics

In quantum mechanics, observables are determined by pairs of states, while in classical theory they refer to the same state. This is especially evident in matrix mechanics which describes the atom as a twofold infinite, denumerable array of virtual oscillators, where observables are vectors in Hilbert space whose magnitude defines a spectral line intensity, or transition amplitude, and whose direction corresponds to either an absorption or an emission. Although the physical model consists of two ideal particles, the virtual harmonic oscillators, the matrix elements include three field components: the fields of the two ideal particles and a spectral line intensity due to photon superpositions. All three components of the modified Hamiltonian given by Eq. (4) are present but with respect to classical space and time. In nonrelativistic theory, photons are singularities that correspond to pairs of states, whereas conformance with relativity theory requires that the photon be spatially and temporally extended. Fields are localized by exact four-dimensional field boundaries separated from each other in space-time, while in nonrelativistic quantum mechanics, field boundaries do not exist.

Because Heisenberg's uncertainty relations use a continuous time parameter, they are only valid when events are defined with respect to specific observers, but not in general for all observers. When interpreted according to Eq. (2) by a fully relativistic theory, we conclude that indeterminacy is due to measurements performed with a non-singular, spatially and temporally extended probe, the photon. This may be compared to the case in classical mechanics of measurements that are performed with a coarsely defined standard. In quantum mechanics, the standard of measurement is the photon, and no matter how high its energy, it cannot be used to localize a point particle more precisely than its wavelength. On the other hand, localizations in atomic clocks occur four-dimensionally with respect to *both* field boundaries, so they occur without measurable error.

4.4 Wave mechanics

In the wave mechanical interpretation of quantum mechanics, field boundaries are not specified. Nevertheless fields are a part of wave functions, and field boundaries must be included in a fully relativistic theory of electrodynamics. To satisfy that requirement, we interpret the wave function $\Psi\left(\vec{r},t\right)$ as combining an electron or other particle, and a force $\varepsilon\left(\vec{r},t\right)$, that are separated by field boundaries. A wave function composed in this manner as a composite of two physical components may be used to describe the interaction of particles in both bound and free states. Whereas in classical theory forces are three-dimensional vectors with direction and magnitude, in a fully relativistic theory they are four-dimensional and symmetric in the coordinates. They have orientation in space-time, but not direction, with magnitude determined by the instantaneous separation of field boundaries according to (1). Thus force is the continuous application of a discrete form. All interactions of electrodynamics may be conceived of in this way in terms of fields and their boundaries.

5. Discussion

Although the wave function contains all that can be known about a particle, the preceding fully relativistic interpretation of atomic structure indicates the presence of internal characteristics that are *in principle* unknown to observers. The field model, described by (1) and confirmed experimentally by slow or stopped light phenomena, includes internal processes in its description of the wave function that are temporarily restricted from external expression due to field boundaries. The characteristics cannot be accessed because the fields vanish at the field boundaries. Due to the unobserved processes, quantum theory predicts the occurrence of instantaneous action-at-a-distance events such as the collapse of the wave function and other macroscopic phenomena that exist outside of our consciousness. However, if wave functions are interpreted in a fully relativistic theory, we conclude that these phenomena are only unusual when interpreted in abstract space with respect to continuous time parameters.

The detection events that form the basis of optical theory are due to energy emissions that occur at singular points in time and are referred to as "photons" due to their discrete nature. If energy absorption evolves according to Eq. (2), as the integration of a Lagrangian density over a region of space-time, then excitation is a continuous process that results from field superposition during the discrete time period τ. Therefore in a fully relativistic theory, interference effects are due to the instantaneous reinforcement and cancelation of superposed photons of the type described in (2), and the statistical nature of quantum mechanics that is observed in experiments such as double-slit interference is due to time averages. A fully relativistic optical theory will account for interference effects as they evolve in real space and time.

6. Conclusion

It has long been asserted that classical physics is inadequate for describing quantum mechanical phenomena. Consequently experimental results are explained by introducing complementarity and the correspondence principle. However, the problem is not that classical theory is deficient, but it is the insistence on using singularities in a nonrelativistic theory. If the photon's fields are singular, wave and particle properties seem to appear out of nowhere, and experimental results have an intrinsically defined uncertainty. But if the photon is instead described as a localization of fields, uncertainty and duality are accounted for by physical characteristics, fields and field boundaries, and complementarity has a classically derived meaning. A similar explanation is possible for the correspondence principle which specifies the point where a two-particle classical system must be replaced by a three-particle quantum mechanical system to explain what is observed. It may seem to be an acceptable practice to describe particles as singularities propagating and interacting continuously in time, but in a fully relativistic theory the photon cannot be singular. Rather it is a four-dimensional localization of fields defined symmetrically in space-time that determines electron behavior in bound states and also in free space by means of four-dimensional forces.

The mathematical framework surrounding quantum mechanics is precisely the type of description that is expected when a particle of zero mass is absorbed by a two-particle system. The particle properties of the photon are overwhelmed by the other two such that it is impossible to distinguish it independently of them. Sometimes the influence of its continuous properties is more evident (wave mechanics); at other times its discrete properties are prominent (matrix mechanics); and in path integral formulations, the exact field boundaries of bound states are manifested. Each of the three formulations of nonrelativistic quantum mechanics provides a unique perspective to atomic structure by emphasizing a different physical aspect of the three field sources. This may be compared to the simpler three-dimensional practice in architecture of providing three visual perspectives to a building. Each one provides a partial view, and when taken together they give an improved understanding of the structure as a whole. The "whole" of quantum mechanics is given of course by Lagrangian quantum mechanics.

Author details

Richard Oldani
Illinois Institute of Technology, Clymer, NY, USA

*Address all correspondence to: oldani@juno.com

References

[1] Dirac PAM. The quantum theory of the emission and absorption of radiation. Proceedings of the Royal Society A. 1927;**114**:243

[2] Jackson JD. Classical Electrodynamics. 3rd ed. Wiley; 1999 pp. 13 ff, 248 ff

[3] Franken PA, Hill AC, Peters CW, Weinreich G. Generation of optical harmonics. Physical Review Letters. 1961;**7**:118

[4] Verma VP et al. Frequency doubling in scattering. Journal of Physics B: Atomic, Molecular and Optical Physics. 1988;**21**:1367

[5] Dirac PAM. The Lagrangian in quantum mechanics. Physikalische Zeitschrift Sow. 1933;**3**:1

[6] Feynman RP. Space-time approach to quantum electrodynamics. Physical Review. 1949;**76**:769

[7] Hinkley N, Sherman JA, Phillips NB, Schioppo M, Lemke ND, Beloy K, et al. 2013. An atomic clock with 10–18 instability. ArXiv:1305.5869

[8] Heinze G, Hubrich C, Halfmann T. Stopped light and image storage by electromagnetically induced transparency up to the regime of one minute. Physical Review Letters. 2013; **111**:033601

[9] Feynman RP. Space-time approach to non-relativistic quantum mechanics. Reviews of Modern Physics. 1948;**20**: 367

Chapter 3

Twin Boundary in hcp Crystals: Quantum and Thermal Behavior

Victor A. Lykah and Eugen S. Syrkin

Abstract

The 180° twin boundary (TB) (stacking fault) is investigated in the hexagonal close-packed (hcp) light materials. It is shown that atomic symmetry inside the twin boundary is lower than in hcp phase due to interatomic interaction between neighbors. In the case of quantum or thermal behavior, for the isosurfaces, an initial spherical form (in hcp phase) transforms into ellipsoid (in the boundary). We introduce the isosurface deformation parameter. The self-consistent description is developed to estimate the parameters of the thermodynamic potential, and the models of hard spheres and ellipsoids are used. It is shown that the quantum or thermal behavior of the boundary atoms causes the following effects: (i) the increase of degree of overlap of the atomic wave functions or trajectories within the twin boundaries, (ii) the increase of diffusion inside the twin boundaries, and (iii) the decrease of energy and broadening of the quantum boundary in comparison with the classical case.

Keywords: solid helium, twin boundary, stacking fault, hard ellipsoids, quantum diffusion
PACS numbers: 61.72.Mm, 61.72.Nn, 64.75.Gh, 66.30.Ma

1. Introduction

Helium crystals have unique quantum properties and are useful for the investigation of dynamic and kinetic behavior of atomic crystals [1]. In experiments, the structure phase transition between body-centered cubic (bcc) and hcp phase was found for both metals and solid ^{4}He [2, 3]. The coherent phase boundary (PB) and twin boundary (TB) or stacking fault (SF) was investigated theoretically in the frame of one order parameter (OP) model [4, 5]. The two-OP theory of PB [6, 7] was developed on the basis of the Burgers mechanism. In work [5] we proposed the three-OP theory that combines Sanati [6, 7] and Kaschenko [8] treatments; so we take into account the changes of volume and pressure under the phase transition. The three-OP and one-OP descriptions of PB and TB are uniquely related. In different models of coherent bcc-hcp boundary, the local oscillations spectra of OP in ^{4}He were investigated in [9].

In the experiment [10], a glass formation under deformation of solid helium was investigated. Usually the deformation of crystals generates the different defects [11], including stacking faults. In the nuclear magnetic resonance (NMR) experiments [12], the great role of the interface in increasing the quantum diffusion was found. In work [13] for single hcp crystal ^{4}He, the stacking fault energy was measured.

The present work is devoted to the development of the self-consistent description of quantum behavior of ^{4}He atoms in twin boundary proposed in work [14]. We apply this treatment to quantum and thermal description of twin boundary in some metals.

2. Model of the twin boundary

In the hcp phase of crystal ^{4}He, we consider the twin boundary under transition from the close packing layers ABAB ... (see **Figure 1a,b**) to the close packaging ACAC .... The twin boundary (TB) corresponds to stacking faults (SF). The atomic plane A creates different positions (potential wells) B and C for neighbor layers (see **Figure 1a,b**).

The twin boundary was researched in works [4, 5] where the triple-well thermodynamic potential was used. Far from the bcc-hcp transition, the double-well free energy can be applied:

$$F(\xi) = \int \left[\frac{\alpha}{2}\left(\frac{d\xi}{dz}\right)^2 + \frac{k_4\xi^4}{4} - \frac{k_2\xi^2}{2}\right] dv, \tag{1}$$

where the integration is over the volume v, square brackets contain the volume energy density, z is a coordinate in the direction of heterogeneity, α is a dispersion parameter responsible for the boundary width, and phenomenological parameters k_4, k_2 are positive. In hexagonal lattice, ξ is the order parameter which means the relative displacement of the atomic layers between positions B and C (see **Figure 1a,b**). For the homogeneous part of the free energy Eq. (1), the maximum and minima positions are

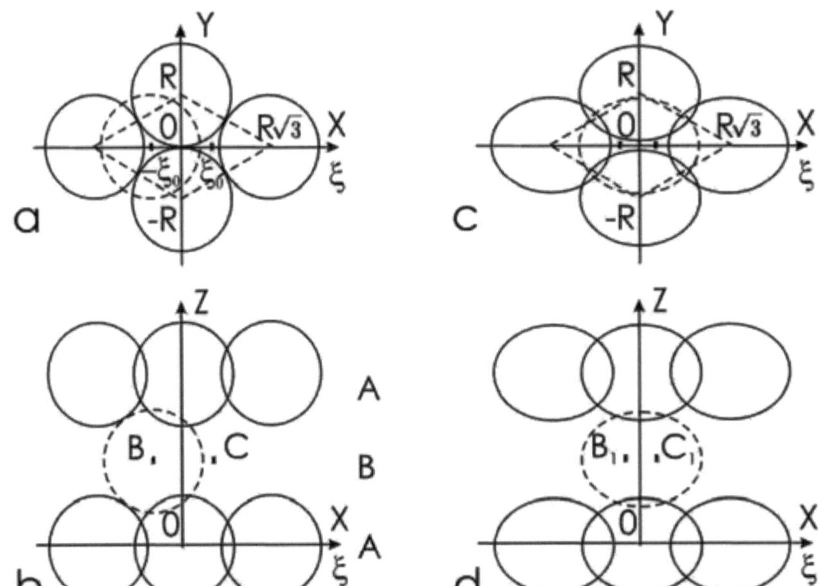

Figure 1.
(a, b) The close pack of the atomic layers (0001) ABAB... for hcp phase. Layer A is shown by solid lines, and layer B is shown by dotted lines. (a) The view perpendicular to the layers. (b) The view along the layers. Points B and C are atomic equilibrium positions in corresponding layer. (c, d) The change in the close packing of the atomic ellipsoids inside TB under quantum effects is accounted.

$$\xi_{max} = 0; \qquad F(\xi_{max}) = 0;$$
$$\xi_{min} = \pm\xi_0; \quad F(\xi_{min}) = -V\frac{k_4\xi_0^4}{4}; \qquad \xi_0 = \sqrt{\frac{k_2}{k_4}}. \tag{2}$$

where $|\xi_0|$ is the minimum position as displacement between the maximum and minimum positions (B and C in **Figure 1a, b**). The difference between the maximum and minimum energies gives the height h of the potential barrier per unit volume:

$$h = \frac{1}{V}[F(\xi_{max}) - F(\xi_{min})] = \frac{k_4\xi_0^4}{4} = \frac{k_2\xi_0^2}{4}. \tag{3}$$

For further analysis it is convenient to write the free energy Eq. (1) in terms of ξ_0 and h:

$$F(\xi) = \int\left[\frac{\alpha}{2}\left(\frac{d\xi}{dz}\right)^2 + \frac{h}{\xi_0^4}\left(\xi^2 - \xi_0^2\right)^2 - h\right]dv. \tag{4}$$

The free energy Eq. (4) gives rise to such one-dimensional inhomogeneity as twin boundary [9, 15] which has shape

$$\xi = \pm\xi_0 \tanh\frac{z}{l_T}; \tag{5}$$

$$l_T = \sqrt{\frac{2\alpha}{k_2}} = \frac{1}{\xi_0}\sqrt{\frac{2\alpha}{k_4}}. \tag{6}$$

where the boundary center is chosen at $z = 0$ and l_T is the characteristic width of the boundary. The shear dependence on coordinate Eq. (5) can be substituted into relation Eq. (4). The surface energy density of the twin boundary is obtained by further integration:

$$W_T = \frac{\sqrt{2\alpha k_2^3}}{3k_4} = \frac{4}{3}l_T h. \tag{7}$$

It is expressed through parameters (α, k_2, k_4) of the microscopic double-well potential or macroscopic parameters (l_T, h). The parameters of the thermodynamic potential Eq. (1) can be transformed into the microscopic ones:

$$k_{21} = \frac{4h_1}{\xi_0^2}; \quad k_{41} = \frac{4h_1}{\xi_0^4};$$
$$h_1 = hv_1;\ k_{21} = k_2 v_1;\ k_{41} = k_4 v_1. \tag{8}$$

Here h_1, k_{21}, and k_{41} are the barrier height Eq. (3), parameters k_2 and k_4 Eq. (1) normalized per unit cell. These equalities are obtained by multiplying h and k_2 or k_4 to the unit cell volume v_1. The characteristic width Eq. (6) $l_T \simeq 1.5nm$ was obtained by molecular dynamic method in [16].

3. Atomic potential in continual description

In hcp lattice, one can find the symmetry axes (along $0z$) of third and sixth orders. In the close-packed layers (x, y), hcp demonstrates isotropic properties of macroscopic tensors [11, 17]. The isotropic macroscopic tensors exist at appropriate relations $c/a = \sqrt{8}/3$ of unit cell sizes [1, 11]. Inside the perfect hcp phase, an atom is in high symmetric (isotropic) potential:

$$U_{is}(\mathbf{r}) = \frac{1}{2}k_{is}\left(x^2 + y^2 + z^2\right). \tag{9}$$

where k_{is} is isotropic rigidity. The harmonic approximation Eq. (9) is satisfied better for heavier inert atoms or light metals; however, the helium crystal has pronouncedly anharmonic atomic potential [18]. Nevertheless in helium crystals, the harmonic approximation is successfully applied [19, 20].

The isotropic rigidity k_{is} can be divided into two contributions: $k_{is} = k_p + k_{pn}$, where k_p is rigidity in the plane and k_{pn} is rigidity from the interaction with the neighbor planes.

Inside the twin boundary, the neighbor layers are shifted from the symmetric positions, and it causes an anisotropic atomic potential. The previous spherical potential is broken. Then inside the twin boundary, the initial isotropic atomic potential transforms into

$$\begin{aligned} U_{an1}(\mathbf{r}) &= U_{is}(y,z) + U_{an1}(x); \\ U_{is}(y,z) &= \frac{1}{2}k_{is}\left(y^2 + z^2\right); \; U_{an1}(x) = U_p(x) + U_{pn}(x,\xi); \\ U_p(x) &= \frac{1}{2}k_p x^2; \; U_{pn}(x,\xi) = \frac{k_{41}(\xi - x)^4}{4} - \frac{k_{21}(\xi - x)^2}{2}. \end{aligned} \tag{10}$$

where the isotropic potential Eq. (9) splits into two terms. The first term $U_{an1}(x)$ is an anisotropic and nonlinear part of the potential in the shift direction Ox. The second term $U_{is}(y,z)$ is the rest of the isotropic part which is perpendicular to the shift direction. Further, the potential $U_{an1}(x)$ is divided too into $U_p(x)$, the isotropic part, and $U_{pn}(x,\xi)$, the anisotropic one from the neighbor atomic planes. The last turn depends on the layer shift ξ and the small deviation x. Therefore, only term $U_{pn}(x,\xi)$ changes inside TB which is shown in **Figure 2**. The analysis (see [14]) of the term $U_{pn}(x,\xi)$ allows to write the anisotropic atomic potential Eq. (10) in the following simple form:

$$\begin{aligned} U_{an1}(\mathbf{r},\xi) &\simeq U_0(\xi) + c(\xi)x + \frac{1}{2}k_b(\xi)x^2 + \frac{1}{2}k_{is}\left(y^2 + z^2\right); \\ k_b(\xi) &= k_p + k_{pn}(\xi) = k_{is} + 3k_{21}\left(\frac{\xi^2}{\xi_0^2} - 1\right); \; k_{pn}(\xi) = +k_{21}\left(3\frac{\xi^2}{\xi_0^2} - 1\right). \end{aligned} \tag{11}$$

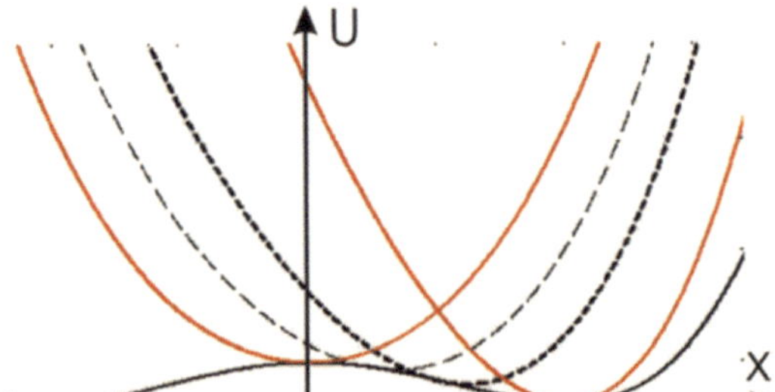

Figure 2.
Smooth changed parts of the potential in dependence on the coordinates ξ and x according to Eqs. (10) and (11): $U_{pn}(0,\xi)$ is a lower double-well curve and $U_{an1}(x,\xi)$ is a set of parabolas.

where $k_b(\xi)$ is rigidity coefficient inside TB, $U_0(\xi)$ is a varied bottom level, and $c(\xi)x$ is the linear part. In the limit points $\xi = \pm\xi_0$, Eq. (11) transforms into isotropic hcp phase Eq. (9) with $k_{pn}(\xi_0) = +2k_{21}$. Inside TB $\xi = 0$, the rigidity takes value $k_{pn}(0) = -k_{21}$. Thus, the rigidity coefficients in phase (k_{is}) and in the middle of TB $(k_b(0))$ are represented by the rigidity coefficients inside the plane (k_p) with two adjacent (k_{21}) planes:

$$k_{is} = k_p + k_{pn}(\xi_0) = k_p + 2k_{21}; \tag{12}$$

$$k_b(0) = k_p + k_{pn}(0) = k_p - k_{21}. \tag{13}$$

Inside the boundary the potential is considerably softer in direction Ox because of $k_b(\xi) < k_{is}$ (see **Figure 2**). The difference in these rigidity coefficients is too high $k_{is} - k_b(0) = 3k_{21}$. For further analysis, we need especially the quadric form in Eq. (11).

The ratio of the rigidity coefficients in the relation Eq. (10) can be related to the ratio of the elastic modules which are shown in **Table 1**. The macroscopic tensor components $C_{11}C_{33}$ describe the longitudinal deformation along the axes $0x$ and $0z$, respectively. In solid ^{4}He, the ratio of the elastic modulus $C_{33}/C_{11} = 1.37$ gives anisotropy of the rigidity coefficients k_{elz}/k_{is} in the basal plane and axis $0z$. Uniaxial compression-tension in the basal plane of $0xy$ corresponds to the elastic modulus of C_{11} and atomic rigidity coefficient k_{is}. The shuffle of the basal planes in an arbitrary direction corresponds to elastic modulus C_{44} and atomic rigidity coefficients $2k_{21}$. Therefore, we have the following inequality:

$$\frac{2k_{21}}{k_{is}} = \frac{2k_{21}}{k_p + 2k_{21}} \lesssim \frac{C_{44}}{C_{11}}. \tag{14}$$

4. The atomic potential and hard sphere model in hcp phase

The geometry of the hcp lattice is shown in **Figure 1a**. In the hard sphere model for the hcp plane A (see **Figure 1a,b**), the coordinates of atomic centers are

$$(0, \pm R_0, 0); \quad \left(\pm R_0\sqrt{3}, 0, 0\right); \tag{15}$$

where R_0 is the atomic radius, x is a coordinate along the shift direction of the atomic plane B, z is a coordinate along the direction perpendicular to the atomic plane, and y is a coordinate along the atomic plane perpendicular to the shift direction. (0,0,0) is the touch point of the spheres in plane A. Then the sphere

Element	C_{11}, GPa	C_{33}, GPa	C_{13}, GPa	C_{44}, GPa
^{4}He [21][a]	$4{:}05{\cdot}10^{-2}$	$5{:}54{\cdot}10^{-2}$	$1{:}05{\cdot}10^{-2}$	$1{:}24{\cdot}10^{-2}$
^{7}Li [22][b]	14.2	—	—	10.7
^{9}Be [23][b]	292	349	6	163
^{24}Mg [23][b]	59.3	61.5	21.4	16.4

[a]*The elastic moduli of hcp ^{4}He are found at T $\sim$ 1K and molar volume 20.97·10^{-6}m^3/mo1 [21].*
[b]*At room temperature.*

Table 1.
The experimental values of the elastic moduli of some hcp materials in the notation of Voigt C_{ik} following [11].

centers of the shifting atomic plane B can move over the following four spherical surfaces:

$$x^2 + (y \pm R_0)^2 + z^2 = (2R_0)^2; \\ (x \pm R_0\sqrt{3})^2 + y^2 + z^2 = (2R_0)^2; \tag{16}$$

The equilibrium points for the atom of the shifting neighbor atomic plane B can be found from the geometry of the system (Eq. (16) at y = 0):

$$x_{Re} = \pm R_0 \frac{1}{\sqrt{3}}; \quad y_{Re} = 0; \quad z_{Re} = R_0\sqrt{\frac{8}{3}}. \tag{17}$$

Signs – and + in x_{Re} describe positions B and C in plane B, respectively. From the first Eq. (16), the saddle point coordinates for an atom of plane B are

$$x_{Rs} = 0; \quad y_{Rs} = 0; \quad z_{Rs} = R_0\sqrt{3}. \tag{18}$$

For the hard sphere model, the microscopic parameters ξ_0, h_1 are

$$\xi_{0R} = |x_{Re}|; \quad h_{1-R} = \frac{1}{2} g k_{is} (z_{Rs} - z_{Re})^2 \tag{19}$$

where h_{1-R} is the potential barrier between B to C position (see **Figure 1**). Coefficient $g \sim 1$ evaluates the quasielastic energy. In the middle of TB, the neighbor number is 4, which is less than 6 once inside the phase. This is a microscopic reason for the quasielastic energy behavior.

For the hard sphere model, the substitution of relations (19) into Eqs. (3) and (8) gives the parameters of the microscopic interatomic potential:

$$k_{21-R} = \frac{4h_{1-R}}{\xi_{0R}^2}; \quad k_{41-R} = \frac{4h_{1-R}}{\xi_{0R}^4}. \tag{20}$$

For comparison, Eq. (11) allows us to find the rigidity coefficients in the phase k_{is} and in the middle of the boundary $k_b(0)$.

5. Quantum atomic spheres and ellipsoids in hcp phase and in the twin boundary

Inside the perfect hcp phase, a ^{4}He atom is in highly symmetric potential of neighbor atoms. In isotropic harmonic approximation [19, 20], the atomic potential can be presented as [24]

$$U_{is}(\mathbf{r}) = \frac{1}{2} m\omega^2 \mathbf{r}^2; \quad \lambda = \frac{m\omega}{\hbar}. \tag{21}$$

where m, $\mathbf{r}$ and λ are mass, radius vector of ^{4}He atom and parameter of the quantum oscillator. The potential Eq. (12) gives $m\omega^2 = k_{is}$; $\lambda^2 = k_{is} m/\hbar^2$.

The Schrodinger equation splits into three equivalent independent equations with the constant $k^2 = k_X^2 + k_y^2 + k_z^2 = 2mW/\hbar^2$ where k_i are wave numbers. The ground state solution [24] has total zero-point energy $W_{0is} = \frac{3}{2}\hbar\omega$. In isotropic harmonic approximation, a distribution of probability density $\rho = |\psi(x, y, z)|^2$ of helium atom has spherical symmetry. Hence, the equation of probability isosurface (sphere of radius R) is

$$x^2+y^2+z^2=R^2;\quad R^2=\frac{N_{\rho is}}{\lambda};\quad N_{pis}=\ln\sqrt{\frac{\lambda^3}{\rho^2\pi^3}}. \tag{22}$$

The probability density at a distance of R_0 that equals to the radius of the atom in the hcp phase (half the distance between the centers of neighboring atoms in the crystal) is

$$\rho_0=\sqrt{\frac{\lambda^3}{\pi^3}}\exp\left(-\kappa_0^2\right)\;;\quad N_{\rho is}(R_0)\equiv\kappa_0^2=R_0^2\lambda. \tag{23}$$

Here we have introduced the dimensionless parameter κ_0 that is important for further consideration. This parameter is proportional to the atomic radius $\kappa_0\sim R_0$ and depends on the isotropic rigidity of the atomic lattice $\kappa_0\sim\lambda^{1/2}\sim k_{is}^{1/4}$. In respect to a huge change in the volume of solid helium [1], the parameter κ_0 can vary widely.

An anisotropic harmonic potential can be written as [24]

$$\begin{aligned}&U_{anis}(\mathbf{r})=\frac{1}{2}m\left(\omega_X^2x^2+\omega_y^2y^2+\omega_z^2z^2\right)\;;\\&\lambda_X=\frac{m\omega_X}{\hbar};\quad\lambda_y=\frac{m\omega_3}{\hbar};\quad\lambda_z=\frac{m\omega_z}{\hbar}.\end{aligned} \tag{24}$$

The parameters λ_i are related to the rigidity coefficients:

$$\lambda_X^2=\frac{m}{\hbar^2}k_{xel};\quad\lambda_y^2=\frac{m}{\hbar^2}k_{yel};\quad\lambda_z^2=\frac{m}{\hbar^2}k_{zel}. \tag{25}$$

In the hcp phase, an anisotropic harmonic approximation is more adequate. Then the rigidity coefficients satisfy inequality $k_{xel}=k_{yel}=k_{is}<k_{zel}$. If we use isotropic harmonic approximation in the hcp phase, then inside of the twin boundary, an atom ^{4}He is in a uniaxial potential of neighboring atoms of Eq. (13): $k_{xel}=k_b\le k_{yel}=k_{zel}=k_{is}$.

The equation splits also into three independent equations with known solutions [24]. Inside TB for the ground state, the distribution of the probability density of the helium atom loses its spherical symmetry. The probability isosurface is ellipsoid with semiaxes $a\ge b\ge c$:

$$a^2=\frac{N_\rho}{\lambda_X};\quad b^2=\frac{N_\rho}{\lambda_0};\quad c^2=\frac{N_\rho}{\lambda_z};\quad N_p=\ln\sqrt{\frac{\lambda_X\lambda_0\lambda_z}{\rho^2\pi^3}}. \tag{26}$$

Parameter N_ρ describes the probability density. If the probability density equals ρ_0 at the atomic radius R_0 in the hcp phase Eq. (23), then $N(\rho)$ takes the following value:

$$N_{\rho_0}=\kappa_0^2+\ln\sqrt{\frac{\lambda_X\lambda_0\lambda_z}{\lambda^3}}. \tag{27}$$

Thus, the relations Eqs. (26) and (27) describe the probability density isosurfaces to find an atom in the anisotropic case. On appropriate limit $\lambda_i=\lambda$, these relations describe the isotropic case.

6. Classic atomic thermal spheres and ellipsoids in hcp phase and the twin boundary

Inside the perfect hcp phase, an atom is positioned in highly symmetric potential of neighbor atoms (see Eq. (9)) and quantum analogue Eq. (21). For any direction, the average thermal energy of an atom is $k_B T/2$ where k_B and T are the Boltzmann constant and temperature.

In isotropic harmonic approximation Eq. (9), the average thermal energy of an atom corresponds to the average potential isosurface (sphere of radius R):

$$x^2 + y^2 + z^2 = R^2; \quad R^2 = \frac{k_B T}{k_{is}}; \tag{28}$$

The general anisotropic potential has form Eq. (10). In anisotropic harmonic case, the potential can be written with corresponding rigidity coefficients as (compare with Eq. (24))

$$\begin{aligned} U_{anis}(\mathbf{r}) &= \frac{1}{2}\left(k_X^2 x^2 + k_y^2 y^2 + k_z^2 z^2\right); \\ k_X = k_{xel}; \quad k_y &= k_{yel}; \quad k_z = k_{zel}. \end{aligned} \tag{29}$$

Then inside of the twin boundary, an atom is in the uniaxial potential of neighboring atoms Eq. (13): $k_{xel} = k_b \le k_{yel} = k_{zel} = k_{is}$.

The motion equation splits also into three independent equivalent equations. The equation of the potential isosurface is ellipsoid with semiaxes $a \ge b \ge c$ (compare with Eq. (26)):

$$\begin{aligned} &\frac{x^2}{a^2} + \frac{y^2}{b^2} + \frac{z^2}{c^2} = 1; \\ a^2 = \frac{k_B T}{k_X}; \quad &b^2 = \frac{k_B T}{k_y}; \quad c^2 = \frac{k_B T}{k_z}. \end{aligned} \tag{30}$$

Thus, the relation Eq. (30) describes the atomic potential isosurfaces in the anisotropic case, i.e., inside TB. In the limit case $k_i = k_{is}$, it corresponds to the isotropic case, i.e., hcp phase Eq. (28). The thermal potential isosurfaces (ellipsoids) have to be in order less than the quantum atomic spheres and ellipsoids normalized at R_0. We emphasize that in this section the average thermal motion of atoms was considered.

7. The self-consistent description of the twin boundary

The **classic description** of TB uses two coefficients of the thermodynamic potential Eq. (1):

$$k_{21} = const; \quad k_{41} = const; \quad \text{or } h_1 = const; \quad \xi_0 = const. \tag{31}$$

They can be corresponded to the hard sphere model (see Eqs. (15)–(20)).

The quantum and thermal description of TB is self-consistent, i.e., the parameters Eq. (31) are varied as a function of some parameter q that, in its turn, is a function of these parameters:

$$h_1 = h_1(q); \quad \xi_0 = \xi_0(q); \quad q = q(h_1, \ \xi_0). \tag{32}$$

Let us introduce the **isosurface deformation parameter** q as a geometric factor which describes the deformation of the atomic sphere Eqs. (22) and (28) into the one-axis ellipsoid Eqs. (26) and (30):

$$q = 1 - \frac{c^2}{a^2} \equiv \varepsilon^2; \quad 0 \leq \varepsilon^2 \leq 1. \tag{33}$$

where ε is the eccentricity of the ellipse. Earlier in the paper [14], we introduced the quantum deformation parameter q_q. Here we generalize the parameter q_q to the cases of either quantum or thermal motion of an atom and introduce the isosurface deformation parameter q.

Now we present the self-consistent scheme of description for the twin boundary.

(0) Zero approximation. An atom is a hard classic sphere Eq. (32) or quantum isotropic oscillator:

$$R_0 = a = b = c; \quad \rho = \rho_0; \quad q = 0. \tag{34}$$

(1) The first approximation. An atom is considered as a quantum anisotropic uniaxis oscillator. The potential Eq. (10) has been obtained in zero approximation. In the general case, the ellipsoid parameters and the isosurface deformation parameter are described by Eqs. (26), (27), and (33), respectively. The long ellipsoids axis is oriented along the shift direction $0x$:

$$b_1 = c_1 < a_1; \quad \rho = \rho_0; \quad \varepsilon_{c1}^2 = q_1 = 1 - \frac{c_1^2}{a_1^2} > 0. \tag{35}$$

The further variations of **parameters** Eq. (31) can be obtained in the hard ellipsoid model. The hard ellipsoids have the isosurfaces with the same probability density ρ_0 as the hard spheres in the hcp phase, and the isosurface deformation parameter can be obtained. For a vacancy, the nearest neighbors form similar ellipsoids [25].

(2) The second approximation. An atom is considered as an anisotropic three-axis oscillator (the isosurface is three-axis ellipsoid). The first approximation gives the rigidity coefficients of the potential. Different ellipses are formed in the planes ab and ac, and their eccentricities equal

$$b_2 \neq c_2 < a_2; \quad \varepsilon_{b2}^2 = q_{b2} = 1 - \frac{b_2^2}{a_2^2} > 0; \quad \varepsilon_{c2}^2 = q_{c2} = 1 - \frac{c_2^2}{a_2^2} > 0. \tag{36}$$

Now all three axes of the atomic ellipsoid are different. The softest potential and the longest axis a_2 are still oriented along the shift direction. The hard ellipsoid model Eq. (35) is used to obtain a new local atomic potential and a new ellipsoid shape.

(i) The third and further i^{th} steps qualitatively replicate the previous steps in the same way. The second and further steps are more cumbersome and complicated.

8. Atom as anisotropic harmonic oscillator in the boundary, one axis

In continual description inside the boundary, we have found a change of the atomic potential Eq. (10) with the corresponding rigidity constants. Therefore, constants λ_i in Eq. (25) take the following forms:

$$\lambda_X = \frac{1}{\hbar}\sqrt{m\left[k_{is} + 3k_{21}\left(\frac{\xi^2}{\xi_0^2} - 1\right)\right]} \le \lambda; \quad \lambda_y = \lambda_z = \lambda = \frac{1}{\hbar}\sqrt{mk_{is}}. \tag{37}$$

Using Eqs. (26) and (27), the atomic isosurface can be described by ellipsoid with semiaxes:

$$a_1^2 = \frac{N_{\rho_0 1}}{\lambda_X}; \quad b_1^2 = c_1^2 = \frac{N_{\rho_0 1}}{\lambda}; \quad N_{\rho_0 1} = \kappa_0^2 + \ln\sqrt{\frac{\lambda_X}{\lambda}}. \tag{38}$$

For fixed λ_y, $\lambda_z = \lambda$ and reduced stiffness coefficient λ_X along axis $0x$, the semiaxes of the ellipsoid change as follows: $a_1 > R_0$; $b_1 = C_1 < R_0$. Then the isosurface deformation parameter q_1 Eq. (35) takes the following dependence on the order parameter ξ and coordinate

$$q_1 = 1 - \sqrt{1 - 3\frac{k_{21}}{k_{is}}\frac{1}{\cosh^2(z/l_T)}}. \tag{39}$$

We obtain the same result for the thermal excitations; however, instead of relation Eq. (37), we use the rigidity constants Eq. (30):

$$k_X = k_{is} + 3k_{21}\left(\frac{\xi^2}{\xi_0^2} - 1\right) \le k_{is}. \tag{40}$$

In **Table 2**, evaluations of different parameters are shown according to **Table 1** and relation Eqs. (14), (39), (42), and (43); the sources are shown in round brackets on top of columns.

In He and Mg (see **Table 2**), the transverse components of the elastic module C_{44} are much smaller than the longitudinal ones C_{11}. Accordingly in these materials, the isosurface deformation parameters in the middle point of TB q_{max} take relatively small value.

In Li and Be (see **Table 2**), the transverse and longitudinal components of the elastic moduli are closer. Hence, in these materials, the parameters q_{max} are considerably greater. Moreover in Li, the parameter q_{max} can reach 1 or even take complex (imaginary root) values. This indicates a possible instability of Li crystal lattice (see further consideration). This, seemingly unexpected, result is quite understandable

Element	$\frac{C_{44}}{C_{11}}$, (Table 1, (14))	$3\frac{k_{21}}{k_{is}}$, Eq. (14)	q_{max}, Eq. (39)	κ_0^2, Eq. (42)	Λ, Eq. (43)
^{4}He	$\simeq 0.306$	≤ 0.46	$\simeq 0.27$	$\simeq 3.77$	$\simeq 0.0663$[a]
^{7}Li[b]	$\simeq 0.75$	≤ 1.13	$\to 1$	$\simeq 151.3$	$\simeq 0.0017$[c]
^{9}Be[b]	$\simeq 0.558$	≤ 0.84	$\simeq 0.60$	$\simeq 127.4$	$\simeq 0.0020$[c]
^{24}Mg[b]	$\simeq 0.277$	≤ 0.41	$\simeq 0.23$	$\simeq 353.4$	$\simeq 0.0007$[c]

[a]*Evaluation of the de Boer parameter* $\Lambda = 0.45$ *for* ^{4}*He at* ~*1 K [2].*

[b]*At room temperature.*

[c]*Evaluation of the de Boer parameter at* ~ *1K (present work).*

For all materials the parameters κ_0^2 *and* Λ *are evaluated with the same* R_0.

Table 2.
Evaluation of the elastic moduli relations, rigidity relations, the isosurface deformation parameter in the middle point of TB q_{max}, *and the de Boer parameter* Λ *of some hcp materials.*

if considerably gentle upper parabolas (stronger interaction between the crystal planes in comparison with in-plane interaction) are taken into account which are shown in **Figure 2**.

In the quantum case, we can evaluate the minimal increase of the exchange integral due to the increase of overlapping wave functions caused by the elliptic deformations [14]:

$$\Delta I = \Delta I_0 = \frac{1}{\cosh^4(z/l_T)};$$
$$\Delta I_0 = \frac{3}{16\sqrt{\pi}} \frac{1}{\sqrt[4]{\kappa_0^5}} \left(\frac{k_{21}}{k_{is}}\right)^2 \exp\left(-\kappa_0^2\right); \tag{41}$$

Increasing overlapping volumes ΔV with high probability can be evaluated by segments of the crossing ellipsoids. Amplitude ΔI_0 depends on two parameters κ_0 and k_{21}/k_{is} only.

In the basal hcp plane, the exchange integral is varied depending on the quantum deformation parameter q; the wave function tails are the most sensitive, especially in the overlapping region. Evaluations Eq. (41) take into account only space changing but not the amplitude one. The amplitude changing can achieve several orders because of exponential dependence. The exchange integral I uniquely defines the diffusion coefficient [26]. In the interphase boundaries in solid helium, NMR experiment [12] shows the quantum diffusion increasing. The interphase and twin boundaries are similar [5]. So for the quantum diffusion case in TB, the predicted and the experimentally observed arising values are closely related. Experiments show thermal diffusion arising at boundaries [11]; the found thermal ellipsoids' deformation qualitatively explains these facts.

Now we can point out conditions when exchange integral Eq. (41) increases. We need minimal κ_0^2 Eq. (42) in exponent Eq. (41). The parameter κ_0 or λ can be defined by Eqs. (23) and (37) and analyzed in dependence on different factors. In [14] using atomic mass m_a and evaluation of atomic radius R_0, the parameter κ_0^2 value was estimated:

$$\kappa_0^2 \simeq \frac{1}{\hbar} R_0^{5/2} \sqrt{\frac{1}{2}\pi m_a E}. \tag{42}$$

where elastic module $E = C_{11}$ is related to the rigidity coefficient $k_{is} \simeq \pi R_0 E/2$. In solid ^4He, the atomic radius R_0 is the soft parameter, especially under low pressure. So, a high value of exchange integral can be achieved. Compressibility is small in metals, first of all, in light ones (lithium, beryllium, magnesium). Minimal rigidity k_{is} gives rise in the exchange integral too. The van der Waals interaction in ^4He is 3–4 orders of magnitude less than in metal (see **Table 1**).

Another way to estimate κ_0^2 is to compare it with de Boer parameter Λ, the fundamental characteristic of quantum crystal. The de Boer parameter gives the probability density to find an atom in the site of a neighboring atom (at distance $a_l = 2R_0$) [26]:

$$\rho(a_l) \sim \exp\left(-\frac{1}{\Lambda}\right) = \exp\left(-\lambda a_l^2\right) \;;\; \kappa_0^2 = \frac{1}{4\Lambda}. \tag{43}$$

The de Boer parameter $\Lambda = 0.45$ for ^4He [26] gives evaluation $\kappa_0^2 \simeq 0.59$. Pressure growing leads to more difficult tunneling of atoms and different κ_0^2 evaluations in Eqs. (42) and (43). Using the data in **Table 1**, for solid ^4He we obtain R_0 [14], $\kappa_0^2 \simeq 3.77$, and $\Lambda \simeq 0.07$ (see **Table 2**).

We can make the following conclusion. The softening of the effective atomic potential is anisotropic inside the twin boundary which increases the exchange integral and tunneling probability in the selected shear direction. As a result the quantum diffusion along the boundary plane increases.

9. The self-consistent correspondence of the potential and the uniaxial hard ellipsoid model

Inside the twin boundary, the arising anisotropic atomic potential transforms an atomic probability isosurface from sphere to ellipsoid. Let us introduce the hard ellipsoid model as analogue of the hard sphere model. Then coefficients' local values for the potential can be found inside TB. We suppose that the twin boundary does not change symmetry and positions of the atomic centers inside a shifting plane. So, the atomic plane A keeps the atomic centers' coordinates Eq. (15) under shifting (see **Figure 1c, d**). In the shifting neighbor atomic plane B, the atomic isosurface equation is defined by Eq. (22). Then for the shifting atomic plane B, the atomic (ellipsoids) center moves over the great ellipsoidal surfaces:

$$\left(\frac{x}{2a_1}\right)^2+\left(\frac{y\pm R_0}{2c_1}\right)^2+\left(\frac{z}{2c_1}\right)^2=1;$$
$$\left(\frac{x\pm R_0\sqrt{3}}{2a_1}\right)^2+\left(\frac{y}{2c_1}\right)^2+\left(\frac{z}{2c_1}\right)^2=1; \tag{44}$$

where the equilibrium and saddle points for an atom are located. Only four ellipsoids with centers $(0,\pm R_0,0)$ and $\left(\pm R_0\sqrt{3},0,0\right)$ are described. Axis $0x$ is directed along the shift (see **Figure 1c, d**).

Relations Eq. (26), (27), and (35) define the ellipsoid's semiaxes as function of R_0, q:

$$a_1^2=\frac{N_p}{\lambda_X};\quad q_1=1-\frac{\lambda_X}{\lambda};$$
$$b_1^2=c_1^2=R_0^2\gamma_1(q_1)\ ;\ \gamma_1(q_1)=1+\frac{1}{\kappa_0^2}\ln\sqrt{1-q_1}. \tag{45}$$

Accounting these relations and condition $y=0$ (see **Figure 1**), we obtain solution for the equation system Eq. (44) and the equilibrium point coordinates for the atom of the plane B. So, in the hard ellipsoid model, we find the microscopic parameters Eqs. (2) and (3) of the atomic potential:

$$\xi_{0-1}=|x_{1-e}|=\xi_0\frac{2-3q}{2(1-q)};\quad h_{1-1}\sim\frac{1}{2}k_{is}(z_{1-s}-z_{1-e})^2=$$
$$=\frac{3h_{1-R}}{\left(3-\sqrt{8}\right)^2}\left[\sqrt{4\gamma_1(q)-1}-\sqrt{4\gamma_1(q)-1-\frac{1}{12}\frac{(2-3q)^2}{1-q}}\right]^2 \tag{46}$$

where h_{1-R} is defined in Eq. (19). These results are valid in the range $0\le q\le 2/3$. At $q_1\to 2/3$ we have $\xi_{0-1}, h_{1-1}\to 0$ and semiaxis relation $a_1/c_1=1/\sqrt{3}$. At $q=2/3$ the hard ellipsoid model needs transition in another state (see [14]). Therefore, inside TB, the change of the atomic wave function leads to the following change of the interaction potential: the equilibrium displacement and the potential barrier height decrease (see **Figure 3**). However, the potential barrier height decreases much faster. The resulting evolution of the potential Eq. (46) is shown in **Figure 4**. Then from Eqs. (3) and (46), the coefficients of the potential are

$$k_{21-1} = \frac{4h_{1-1}}{\xi_{01}^2}; \quad k_{41-1} = \frac{4h_{1-1}}{\xi_{01}^4}. \tag{47}$$

Thus, the coefficients of the potential (1) for the shift in the direction $0x$ reduce $k_{21} > k_{21-1}(q)$ and $k_{41} > k_{41-1}(q)$. It means softening of the potential in the direction of the plane shuffle. The correspondence between the hard ellipsoid model and the atomic microscopic potential Eqs. (4), (8), and (46) is shown in **Figure 3**. Elliptical deformation of the probability isosurface leads to the transformation of the potential energy of the atom in Eq. (10):

$$\begin{gathered} U_{an2}(\mathbf{r}) = U_{an2}(y,z) + U_{p2}(x) + U_{pn2}(x,\xi) \ ; \\ U_{an2}(y,z) = \frac{1}{2}k_{xe2}z^2 + \frac{1}{2}k_{ye2}y^2; \ U_{p2}(x) = \frac{1}{2}k_{p2}x^2; \\ U_{pn2}(x,\xi) = \frac{k_{41-1}(\xi - x)^4}{4} - \frac{k_{21-1}(\xi - x)^2}{2}; \end{gathered} \tag{48}$$

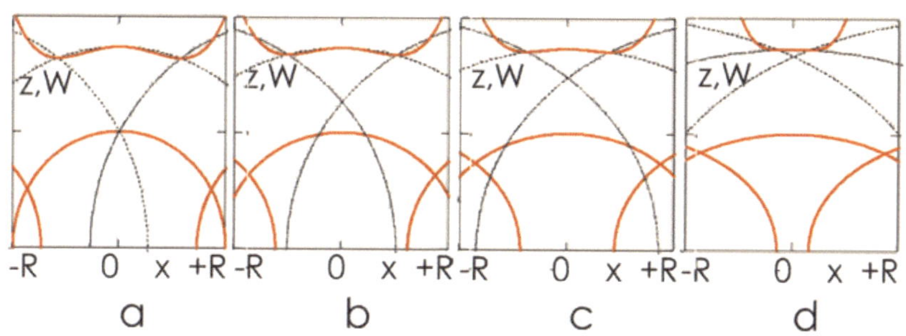

Figure 3.
Comparison of the hard ellipsoids model and the atomic microscopic potential. The red double-well curve shows the potential as a function of ξ_{0-1} and h_{1-1}. Small solid red ellipsoids show atomic isosurfaces at κ_0. Big dot black ellipsoids show the cross sections of the surfaces Eq. (44) at $y = 0$ and quantum parameter values (a) q = 0, (b) q = 0.2, (c) q = 0.4, and (d) q = 0.6.

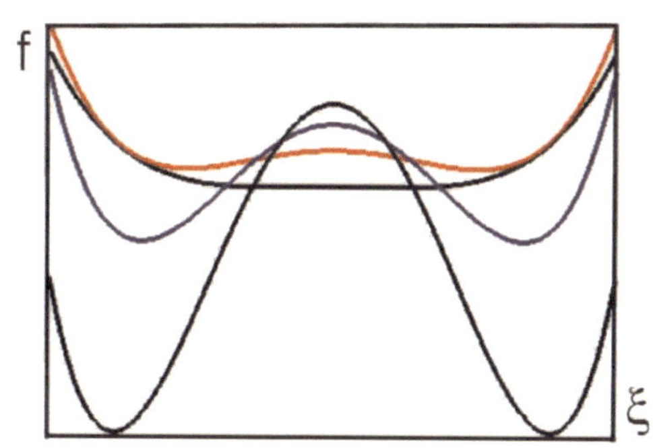

Figure 4.
The cross sections of the potential density according to Eqs. (46), (47), and (39). The quantum boundary has lower potential peak and shorter distance between shallower wells (ξ_{0-1}) as q grows (0, 0.2, 0.4, 0.6). The barrier in the middle of wall (TB) decreases.

All terms are changed in the potential Eq. (48) in comparison with Eq. (10). Isotropy is broken in atomic planes A or B-C due to superposition of the ellipsoids in the shear direction.

For the classical and quantum cases, the free energy density relation Eq. (4) was analyzed analytically in [14]. It was shown that the classical and quantum boundaries have different properties. In particular, from **Figure 5**, it is qualitatively clear why the classical and quantum boundaries have different potential barrier and energy density. In TB both the width and the height of the barrier decrease to zero, according to Eqs. (46) and (47) (see **Figure 4**). In **Figure 3**, they are shown as higher smooth curves. Simultaneously the space width of the boundary $l_T = (1/\xi_0)\sqrt{2\alpha/k_4}$ grows by Eq. (6). The dependence $l_T(q)$ causes further widening of region with $q \to q_{max}$ and a minimal barrier height.

To estimate the energy of the twin boundary (stacking fault) from Eq. (7), we must know the following parameters: α, k_2, k_4 or l_T, h.

The characteristic width (half width) of TB Eq. (6) $l_T \simeq 1.5$ nm was obtained by molecular dynamic method in [16]. We estimate the dispersion parameter α by comparing the differential equations for the transverse sound and shuffling waves:

$$\rho^* \frac{\partial^2 \xi}{\partial t^2} - \alpha \frac{\partial^2 \xi}{\partial z^2} = 0; \quad \frac{\partial^2 u}{\partial t^2} - s^2 \frac{\partial^2 u}{\partial z^2} = 0; \tag{49}$$

where $\rho^* = \rho/2$ is the effective density of the oscillating shuffled subsystem, ρ is the density of helium-4, ξ is the shuffling order parameter, u is macroscopic displacement, and $s = \sqrt{C_{44}/p} \simeq 255$m/s is the transverse sound velocity in the shuffle direction (Oz axis). The velocities of transverse sound and shuffling wave have close values. So the dispersion parameter is $\alpha \simeq C_{44}/2 = 6.2 \cdot 10^6 J/m^3$ where value of module C_{44} is given in **Table 1**.

According to relation Eq. (5), it is possible to estimate the parameter of the thermodynamic potential $k_2 = 2\alpha/l_T^2 \simeq 8.27 \cdot 10^{24}$J/m^5. As follows from Eq. (2) to evaluate the parameter k_4 of the potential, it is necessary to know the maximum displacement of the atom Eq. (17) $\xi_0 = R_0/\sqrt{3} \simeq 1.17 \cdot 10^{-10}$m. Here atomic radius is related to atomic volume: $V_m/N_A \simeq (4/3)\pi R_0^3$. Then $k_4 = 2\alpha/(\xi_0 l_T)^2 = 6.04 \cdot 10^{44}$J/m^7. So, for the classical model of the twin boundary (stacking fault), it is possible to estimate bulk density of the barrier height h and the surface energy density W_T according to Eqs. (3) and (7):

$$h = \frac{k_2 \xi_0^2}{4} \simeq 2.83 \cdot 10^4 J/m^3; \qquad W_T = \frac{4}{3} l_T h \simeq 0.057 mJ/m^2; \tag{50}$$

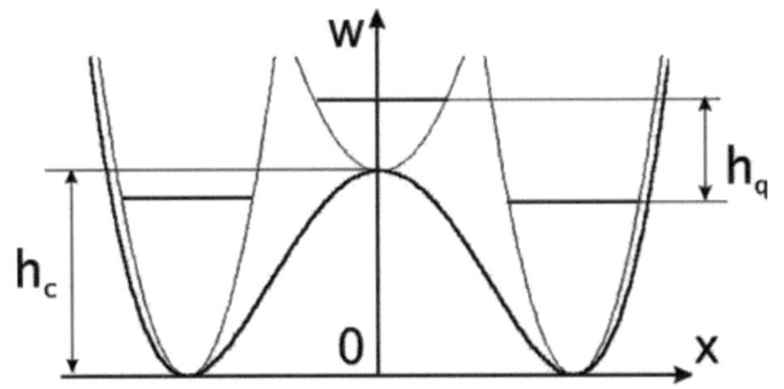

Figure 5.
*The smooth double-well potential according to Eqs. (10) and (11). Instead of a set of parabolas in **Figure 2**, we see only ones at the bottom and the peaks of the potential and their quantum levels. The relationship between the barriers for the atomic displacement in the classical h_c and quantum h_q boundaries is $h_c > h_q$.*

The surface energy density calculated here for the classical model can be compared with the value $W_{SFex} = (0.07 \pm 0.02)\text{mJ/m}^2$ found in the optical experiments at 0.2 K [13].

Above, we have predicted the local reduction of the barrier height h and a local increase in the width l_T of the boundary in the quantum description of the twin boundary (stacking fault). In general for the defect, the surface energy density value $W_T \sim l_{Tef} h_{Tef}$ in Eq. (50) can be close to the classical case. In different experiments and theoretical estimates, a wide variation of the values may be caused by variations of temperatures and pressures.

We have discussed the change of zero vibrations of atoms in the twin boundary (stacking fault) and the related effects. For ^{3}He we can expect the same order of magnitude for all parameters of the twin boundary (stacking fault). The qualitative difference between the pure hcp crystals of isotopes ^{4}He and ^{3}He, apparently, cannot be obtained in the proposed model.

The difference between quantum statistics of the isotopes should address deeper and more delicate quantum properties of the defects. We note briefly below only the most striking manifestation of different statistics and problems arising in this regard.

10. Discussion and conclusion

The quantum self-consistent treatment to twin boundary (stacking faults), proposed in [14] for solid ^{4}He, is developed here for metals and their quantum and thermal description. The relation between discrete models of hard spheres and continuum interatomic potential is used as a sample for a similar relationship in the case of the hard ellipsoid models. As we move deeper into the defect, the transition from one model to another is accomplished.

In the hcp phase, the potential of an atom, created by its neighbors, has spherical symmetry (initial approximation). In the hcp phase, an atom is an isotropic quantum oscillator. In the twin boundary, an atom is an anisotropic quantum oscillator. It is shown that in the twin boundary, the potential of the atom is softer in the direction of shuffle of the atomic planes.

The quantum parameter q_q and its generalization and the isosurface deformation parameter q are introduced. These parameters have simple and visual meaning: q equals to the square of the eccentricity of the cross section of the probability density ellipsoid (or the thermal ellipsoid). We have shown that parameter q is associated with de Boer parameter, the fundamental characteristic of quantum crystal, and anisotropy in the boundary. Evaluations for different materials show that the isosurface deformation parameter q can achieve values $0.2 \div 1$ (see **Table 2**). Meanwhile at $q = 2/3$ the structure instability takes place in the system of the atomic ellipsoids. From this point of view, the properties of TB in lithium are especially interesting because the parameter achieves high value $q \to 1$.

The overlap of the atomic wave functions and the exchange integral value can be described in terms of the quantum parameter q. Inside the twin boundary, the quantum diffusion increases which was observed in the phase boundary (see experiment [12]). The estimation Eq. (50) of the defect energy is in good agreement with experiment [13]. We have shown that the quantum deformation of atoms leads to the space broadening of the twin boundary and to its energy decreasing.

In conclusion we note that local oscillations spectra of the order parameter in different models of coherent bcc-hcp boundary in ^{4}He were investigated in [9]. For small values of the perturbations, dynamical differential equations (reduced to

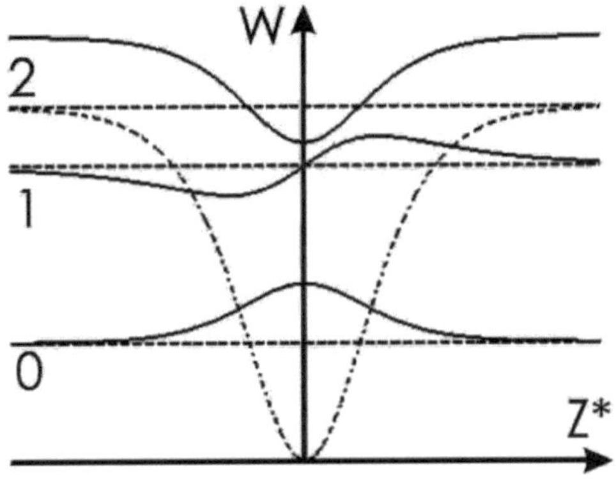

Figure 6.
Local modes of the order parameter at TB [9]. Dash dot line shows the local potential which has local energy levels 0, 1, and 2 (dash). Solid lines show corresponding local oscillations' shape dependence on normalized coordinate z^*.

Schrodinger equations) were obtained and solved. The characteristic frequencies (energy levels) and shape were found and estimated (see **Figure 6**). For the ground state in TB, the local vibration shape can be written as

$$\eta_0(z^*) = \frac{A_0}{\cosh^2 z^*}; \qquad z^* = \frac{z}{l_T}. \tag{51}$$

where A_0 is an amplitude. For the local vibration ground state (51) and for the isosurface deformation parameter q Eq. (39), both shapes coincide qualitatively. In the limit $q < < 1$, both coincide completely. The local vibration of the order parameter describes a correlated motion of the atomic layers in twin boundary. Meanwhile, the quantum and thermal treatments give probabilistic descriptions of the atomic motion. The results (the found smooth arising of the atomic motion amplitude in TB) give evidence that different probabilistic (quantum and thermal) and dynamic methods lead to qualitatively identical features of the atomic basic state inside TB.

Acknowledgements

This research is supported by the FFI National Academy of Sciences of Ukraine, grant 4/18-H, Ministry of Science and Education of Ukraine under the Projects M05486 (0118U002048).

Author details

Victor A. Lykah[1*] and Eugen S. Syrkin[2]

1 Educational-scientific Institute of Physical Engineering National Technical University "Kharkiv Polytechnic Institute", Kharkiv, Ukraine

2 B.I. Verkin Institute for Low Temperature Physics of National Academy of Sciences of Ukraine, Kharkiv, Ukraine

*Address all correspondence to: lykahva@yahoo.com

References

[1] Maradudin AA. In: Freiman YA, Manzhelii VG, editors. Physics of Cryocrystals. AIP Press: New York, Woodbury; 1996. p. 691. DOI: 10.1007/1-56396-537-2

[2] Maekawa M, Okumura Y, Okuda Y. Nucleation of crystals at the bcc-hcp transition line in solid ^{4}He. Physical Review B. 2002;**65**:144525. DOI: 10.1103/PhysRevB.65.144525

[3] Birchenko AP, Vekhov EO, Mikhin NP, Polev AV, Rudavskii EY. Kinetics of the bc-chcp transition in ^{4}He off the melting curve. Low Temperature Physics. 2006;**32**:1118-1112. DOI: 10.1063/1.2400684

[4] Lykah VA, Syrkin ES. Twin boundaries as nuclei of a new phase in body-centered cubic-hexagonal close-packed phase transitions in solid helium. Physics of the Solid State. 2010;**52**: 384-391. DOI: 10. 1134/S10637 83410020265

[5] Lykah VA, Syrkin ES. Theory of the phase and twin boundaries in solid helium and reversibility of the bcc-hcp phase transition. Journal of Low Temperature Physics. 2010;**160**: 179-194. DOI: 10.1007/sl0909-010-0191-6

[6] Sanati M, Saxena A, Lookman T, Alberts RC. Landau free energy for a bcc-hcp reconstructive phase transformation. Physical Review B. 2001;**63**:224114. DOI: 10.1103/PhysRevB.63.224114

[7] Sanati M, Saxena A, Lookman T. Domain wall modeling of bcc to hcp reconstructive phase transformation in early transition metals. Physical Review B. 2001;**64**:092101. DOI: 10.1103/PhysRevB.64.092101

[8] Kaschenko MP, Chaschina VG. Crystal dynamics of the BCC-HCP martensitic transformation: I. controlling wave process. The Physics of Metals and Metallography. 2008;**105**: 537-543. DOI: 10. 1134/S003 1918X08060021

[9] Lykah VA, Syrkin ES. Local oscillations in different models of coherent bcc-hcp boundary in ^{4}He and metals. Physica Status Solidi B. 2011; **248**:1392-1398. DOI: 10. 1002/pssb.201046304

[10] Lisunov AA, Maidanov VA, Rubanskyi VY, Rubets SP, Rudavskii EY, Rybalko AS, et al. Features of the temperature dependence of pressure of solid helium at low temperatures. Low Temperature Physics. 2012;**38**:459-465. DOI: 10.1063/1.4723660

[11] Kittel C. Introduction to Solid State Physics. 8th ed. New York: Wiley; 2004. p. 704

[12] Vekhov EO, Birchenko AP, Mikhin NP, Rudavskii EY. Fast diffusion process in quenched hcp dilute solid ^{3}He-^{4}He mixture. Journal of Low Temperature Physics. 2010;**158**:496-501. DOI: 10.1007/sl0909-009-0022-9

[13] Junes HJ, Alles H, Manninen MS, Parshin AY, Todoshchenko IA. Stacking fault energy in ^{4}He crystals. Journal of Low Temperature Physics. 2008;**153**: 244-249. DOI: 10.1007/sl0909-008-9828-0

[14] Lykah VA, Syrkin ES. Quantum behavior of the twin boundary and the stacking fault in hcp helium crystals. Journal of Low Temperature Physics. 2015;**181**:10-29. DOI: 10.1007/sl0909-015-1328-4

[15] Lines ME, Glass AM. Principles and Application of Ferroelectrics and Related Materials. 2nd ed. Oxford: Clarendon; 2001. p. 683. DOI: 10.1093/acprof:oso/9780198507789.001.0001

[16] Ceperley D. Supersolid: Crystal or plastic? Nature Physics. 2006;**2**:659-660. DOI: 10. 1038/nphys424

[17] Landau LD, Lifshitz EM. Theory of Elasticity. 3rd ed. Vol. 196p. New York: Pergamon; 1986. DOI: 10.1007/978-0-7506-2633-0

[18] Guyer RA. The theory of quantum crystals. In: Seitz F, Tumbull D, Ehrenreich H, editors. Solid State Physics. Vol. 23. New York: Academic Press; 1969. pp. 413-499. DOI: 10.1016/S0081–1947(08)60618–9

[19] Ebner C, Sung CC. Correlation effects in quantum crystals. Physical Review A. 1971;**4**:269-281. DOI: 10.1103/PhysRevA.4.269

[20] Gov N, Polturak E. Local modes, phonons, and mass transport in solid ^{4}He. Physical Review B. 1999;**60**: 1019-1027. DOI: 10.1103/PhysRevB.60.1019

[21] Greywall DS. Elastic constants of hcp ^{4}He. Physical Review B. 1977;**16**: 5127-5128. DOI: 10. 1103/PhysRevB. 16.5127

[22] Day JP, Ruoff AL. The variation of the elastic constants of lithium with temperature and pressure. Physica Status Solidi. 1974;**25**:205-213. DOI: 10. 1002/pssa.2210250118

[23] Tromans D. Elastic anisotropy of hcp metal crystals and polycrystals. International Journal of Research and Reviews in Applied Sciences. 2011;**6**: 462-483. Available from: https://arpapress.com/Volumes/Vol 6Issue4/IJRRAS_6_4_14.pdf

[24] Flugge S. Practical Quantum Mechanics. 2nd ed. Berlin: Springer; 1999. p. 635. DOI: 10.1007/978-3-642-61995-3

[25] Guyer RA, Richardson RC, Zane LI. Excitations in quantum crystals (A survey of NMR experiments in solid helium). Reviews of Modern Physics. 1971;**43**:532-600. DOI: 10.1103/RevModPhys.43.532

[26] Andreev AF. Diffusion in quantum crystals. Soviet Physics Uspekhi. 1976; **19**:137-148. DOI: 10.1070/PU1976v019n02ABEH005133

Chapter 4

Thermoelectric Properties of Oxide Semiconductors

M. Ajaz Un Nabi, Adnan Ali, M. Imran Arshad, Sofia Akber, A. Manzoor, M. Sharif, Kashif Javaid, Nasir Amin, Nadeem Sabir, Khaalid Mahmood, S. Ikram, Sajad Hussain and H. Naeem Ur Rehman

Abstract

In this chapter, we have explored the potential of oxide semiconductors for thermoelectric power generation. Various oxides (Cu_2InO_4, $CuAlO_2$, and Zn_2GeO_4) were grown on Si substrate by thermal evaporation method using tube furnace. After the growth, a representative sample of each oxide was cut into pieces and was annealed at various temperatures from 600 to 800°C in oxygen environment for 1 h using a programmable furnace. The structure of all annealed sample was verified by performing X-ray powder diffraction (XRD) measurements. XRD data suggested that all oxide materials show crystalline behavior at annealing temperature 800°C. XRD results further confirmed that crystal structure of investigated samples improved significantly with annealing because the intensity of oxygen-sensitive (0 0 6) plane was found to be increased with annealing temperature. To investigate the thermoelectric properties of annealed samples, Seebeck effect and Hall effect measurements were performed in the temperature range 25–100°C. It was found that the value of Seebeck coefficient and power factor increased as the annealing temperature increases. Zn_2GeO_4 was found to be a potential thermoelectric material because it has the highest value of Seebeck coefficient and power factor. This highest value is related to the presence of secondary phases in this oxide.

Keywords: oxide semiconductors, thermal evaporation, XRD, Seebeck coefficient, power factor

1. Introduction

Energy has a fundamental importance in the human civilization. Conventional methods are used for the production of energy use oil, gas, and coal. The reservoirs of oil and gas in the world are decreasing, and the burning of oil and gas causes a threat to the environment; therefore, people are searching for cheap methods for the production of clean energy. These renewable energy production methods include photovoltaic, nuclear energy, biogas, wind energy, and thermoelectricity. All these methods have their advantages and disadvantages. For example, solar cells can produce energy during the daylight and also need high technology for the fabrication of solar cells. Nuclear energy production needs nuclear power plants which not only require high cost but also a high risk for the community. Similarly, wind energy can only be produced in the strong windy areas. On the other hand, thermoelectricity is very cheap and an easy method for the clean energy production. Thermoelectricity is based on the very famous Seebeck effect which was invented by Seebeck in 1821 and is stated as.

An emf is induced when a temperature difference is created between two metal junctions. Thermoelectricity needs only temperature difference between two metals; therefore, it is supposed to be the cheapest form of clean energy. It is reported that that 60% of heat produced during cooking process, in industry and in running vehicles, is wasted, but we are able to convert this wasted heat into electricity using thermoelectric power generators; we can save huge amount of money. Furthermore, thermoelectricity has other advantages over other sources of energy such as it has no moving parts, it is environment friendly, no specialized technology is required, and it is less maintenance.

1.1 Physical interpretation

The thermoelectric devices convert thermal energy into electrical energy, and the principle is based on the Seebeck effect invented by Seebeck in 1821. It states that a voltage is induced between two points of metal/semiconductor having a difference of temperature as shown in **Figure 1**. The charge carriers on the hot side can have more energy than the cold side; therefore, they form a potential difference. Suppose dT is the temperature between hot and cold side of sample, therefore according to Seebeck effect.

$$dT = SV \tag{1}$$

where S is the Seebeck coefficient.

Another term frequently used in thermoelectric is the power factor which is defined as.

$$\text{Power factor} = S^2\alpha \tag{2}$$

where α is electrical conductivity.

The performance of thermoelectric material strongly depends upon a unit less quantity called figure of merit:

$$zT = \frac{S^2\sigma T}{\kappa} \tag{3}$$

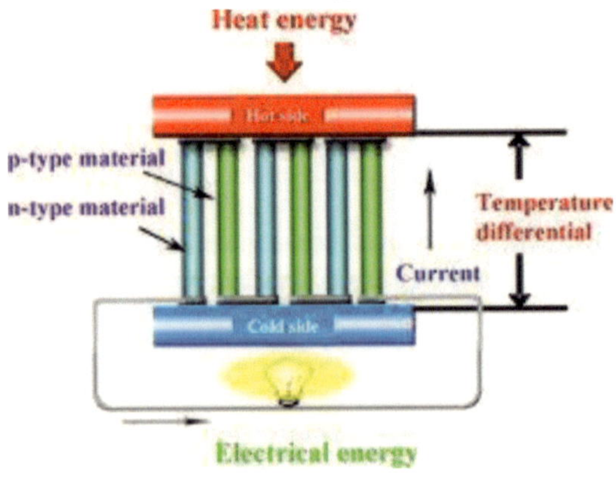

Figure 1.
Schematic diagram of the Seebeck effect.

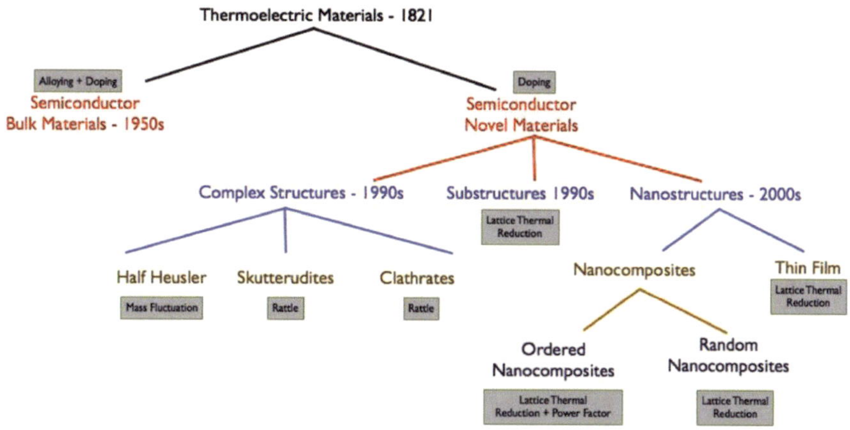

Figure 2.
A brief history of materials used for thermoelectric applications.

This equation shows that for a good thermoelectric material, high Seebeck coefficient, high electrical conductivity, and low thermal conductivity are essential.

The thermoelectric conversion efficiency depends upon a quantity called figure of merit and can be written as [1].

$$ZT = S^2 \alpha/\sigma \quad (4)$$

where S is the Seebeck coefficient, α is the electrical conductivity, and σ is the thermal conductivity.

The figure of merit for a material to be used for practical power generation system should be in the range of 2–3 [2]. But the best reported value of figure of merit for oxide semiconductor is not more than 0.1. **Figure 2** indicates the history of efforts to increase the figure of merit.

Different strategies have been employed to tune and alter the thermoelectric properties. The general techniques are as follows [3]:

1. Optimization using doping techniques

2. Substructuring

3. Nanostructuring

4. Compositing

But for the semiconductors the governing parameters includes the following:

1.2 Band gap

Band structure is a very important parameter to tune the thermoelectric properties of oxide semiconductors. One of the most important methods of band gap control is varying the carrier concentration by doping [4]. But the doping process itself required a high technology that increased the cost of thermoelectric devices very high. So if we tune the carrier concentration by controlling the density of intrinsic defects, this will cut short the cost of the final device. Interestingly, oxide semiconductors have rich the chemistry of intrinsic defects. Oxygen vacancy and zinc interstitials act as intrinsic shallow donors and form electronic states near the conduction band [5].

1.3 Mobility

The mobility of carriers in thermoelectric materials has played a vital role in the tuning of Seebeck coefficient and power factor. The power factor is strongly dependent on the conductivity which has strong dependence on mobility. Therefore, the modulation of mobility to achieve highest value of power factor is necessary. The mobility of the carrier can be controlled by the scattering mechanisms. Two scattering mechanisms, that is, lattice scattering and impurity scattering mechanisms, are very important in oxide semiconductors [6].

1.4 Carrier diffusion

Diffusion is a process of movement of particles from hot junction to cold junction in thermoelectric material. The diffusion of charge carriers has fundamental importance to tune the thermoelectric properties of oxide semiconductors. The diffusion in compound semiconductors is more complex than in elemental semiconductors because of the larger number of possible native point defects that can, in principle, mediate self-diffusion [7, 8]. Oxide semiconductors have high density of intrinsic defects, which in principal affect the diffusion of charge carriers. Therefore, for the effective use of oxide semiconductors for thermoelectric properties, the control of intrinsic defects has fundamental importance and should be studied further. Again, we propose that annealing will be a very effective method of studying the diffusion properties of carriers in oxide semiconductors.

1.5 Thermal expansion coefficient

Thermal expansion is critical, as the devices for high-temperature applications will be subjected to extreme temperature fluctuations. The CTE of TE materials is of critical importance because the shear stress is proportional to the temperature gradient, and the larger the heterogeneity in the thermal expansion coefficient of a material is, the larger is the shear stress that will result [9]. It is also reported that thermal expansion coefficient of semiconductor for low- and high-temperature region is almost the same but different for medium temperature. Therefore, a comprehensive study on the thermal expansion coefficient is still needed to completely understand the mechanism [10].

2. Experimental

In this study, experiment is held under thermal vapor deposition technique using single-stage horizontal glass tube furnace. In this experiment, 99.9% pure magnesium, Zinc, and Copper powders along with Ge, In, and Al powders are used under the ratio 1:1 as source material. This source material is being kept in the center of a glass tube in ceramic boat. The silicon substrate is placed at substrate holder, and the distance between source and the substrate is about 15 cm. The temperature of the furnace is tuned at 950°C for 30 min, whereas the oxygen flow is kept constant at 100 sccm. After the growth of thin film, the substrate of silicon is divided into different pieces for annealing purpose at different temperatures from 600 to 800°C for 30 min.

X-ray diffraction has been performed for the structural analysis of grown thin film. Raman spectroscopy has been also performed to study the rotational and vibrational modes of thin film. Surface morphology is being assessed by the scanning electron microscope (SEM). The most important characterization to calculate

Seebeck coefficient has been performed on the homemade Seebeck system which is based on the four-probe system. And the thickness of the thin film has been measured by the filmtronics technique and it is about.

3. Results and discussion

Figure 3(a–c) represents the XRD patterns of Cu_2InO_4, $CuAlO_2$, and Zn_2GeO_4 thin films annealed at different temperature from 600 to 800°C, respectively. The XRD graph of Cu_2InO_4 thin films in **Figure 3(a)** demonstrated that unannealed and low temperature (600°C) would not be able to make the grown material crystalline due to low thermal energy for bonding. But as we increased the temperature above 600°C, the sample is converted into crystalline structure with preferred orientation (006) at $2\theta = 33.086^{\circ}$ [11, 12]. It is also observed that the intensity of this plane is increased as we further increase the annealing temperature, which suggested that carriers now get enough energy to sit down at a particular position in planes of the crystal. **Figure 3(b)** shows the XRD graph of $CuAlO_2$. The unannealed sample consists of one major phase at $2\theta = 32.05$ which belongs to $CuAlO_2$ (0 0 6) plane [13]. Annealing resulted in the development of new phases at $2\theta = 35.4$, 42.4, and 48.4 related to CuO (1 1 1) and $CuAlO_2$ (1 0 4) and (0 0 9) orientations, respectively. We have observed that (0 0 6) plane has the strongest intensity which is oxygen sensitive; therefore, enhancement of intensity of this plane with annealing temperature is understandable.

Figure 3(c) shows the XRD pattern of grown and annealed samples of Zn_2GeO_4 thin films at various temperatures. The unannealed and annealed samples consist of eight diffraction peaks which are related to Zn_2GeO_4, Si, Au, and ZnO, respectively. The diffraction peak at 25.8, 42.9, 44.7, 58.4, 58.9, and 65° are belonging to Zn_2GeO_4

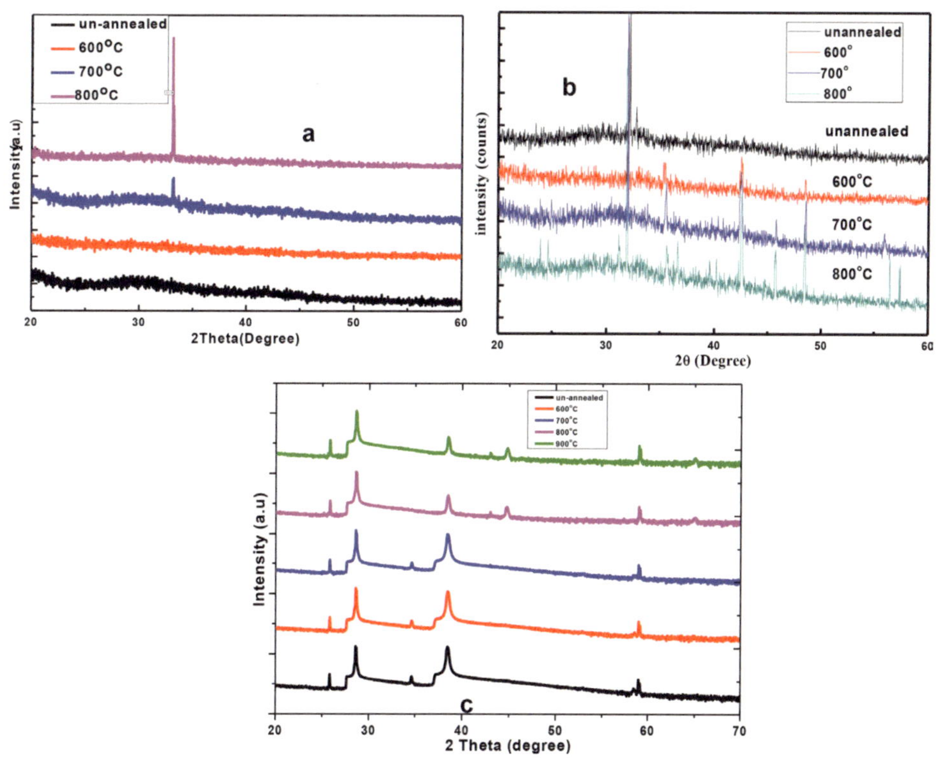

Figure 3.
(a–c) XRD patterns of Cu_2InO_4, $CuAlO_2$, and Zn_2GeO_4, respectively.

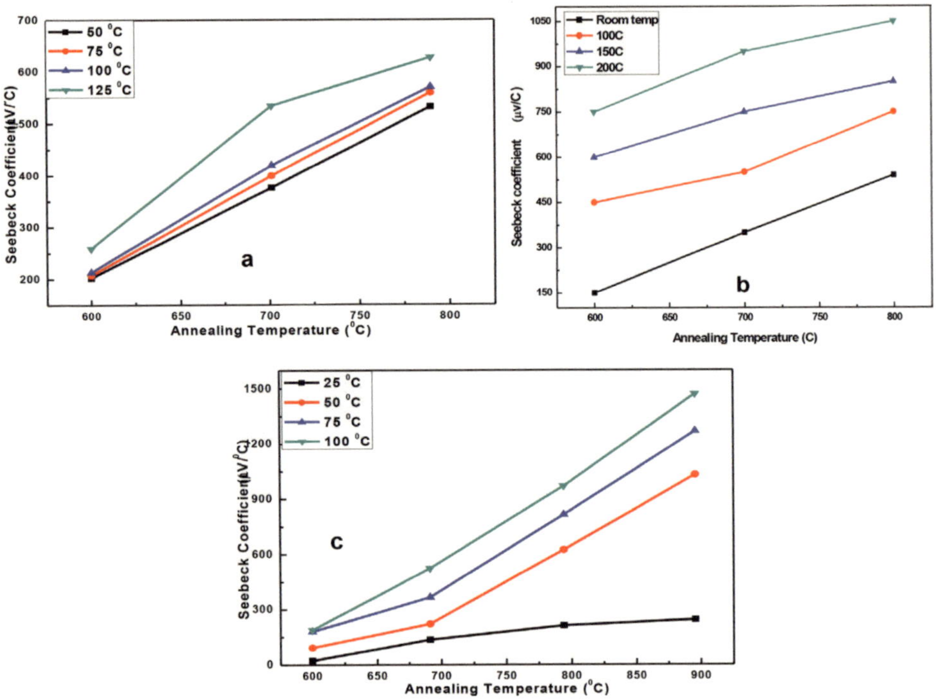

Figure 4.
(a–c) Effect of annealing temperature on the Seebeck coefficient of Cu_2InO_4, $CuAlO_2$, and Zn_2GeO_4, respectively.

which are indexed as (2 2 0), (6 0 0), (0 0 6), (6 3 0), and (7 1 3), respectively [14]. The peak which appears at 28.6° and 38.4° is belonging to (1 1 1) plane of Si and Au [15, 16], respectively, along with peak of ZnO at 34.6° having (0 0 2) plane [17]. It is observed that substrate and pre-deposited gold shows maximum intensity in all the samples which is attributed to the thin porous layer deposited over the substrate. The XRD data also demonstrated that the sample annealed at 900°C has three additional peaks at 45.7° and 65° which are related to the secondary phases of Zn_2GeO_4.

Figure 4(a–c) demonstrated the effect of annealing temperature on the Seebeck coefficient of Cu_2InO_4, $CuAlO_2$, and Zn_2GeO_4 thin films grown by thermal evaporation technique, respectively. All graphs showed that the value of Seebeck coefficient increases as the annealing temperature increased. It is also observed that the value of Seebeck coefficient also increases as the measurement temperature increases from 25 to 100°C. Zn_2GeO_4 has the highest value of the Seebeck coefficient (1470 µV/K) as compared to Cu_2InO_4 and $CuAlO_2$. The observed result can be explained as post-growth annealing enhances the density of oxygen atoms and also provides more thermal energy to Ge atoms which resulted in the creation of GeO-based secondary phases. These newly developed secondary phases act as barrier for charge carriers at the interface of secondary phases. Due to this barrier, the low-energy carriers are filtered out at the interface and caused the enhancement in the Seebeck coefficient. As other two samples have no secondary, therefore have lower value of the Seebeck coefficient as evident by XRD data (**Figure 5**).

To further probe the effect of annealing temperature on the thermoelectric properties of grown oxide semiconductors, we have calculated the power factor using the following formula:

$$P = S^2.\alpha \tag{5}$$

DOI: http://dx.doi.org/10.5772/intechopen.88709

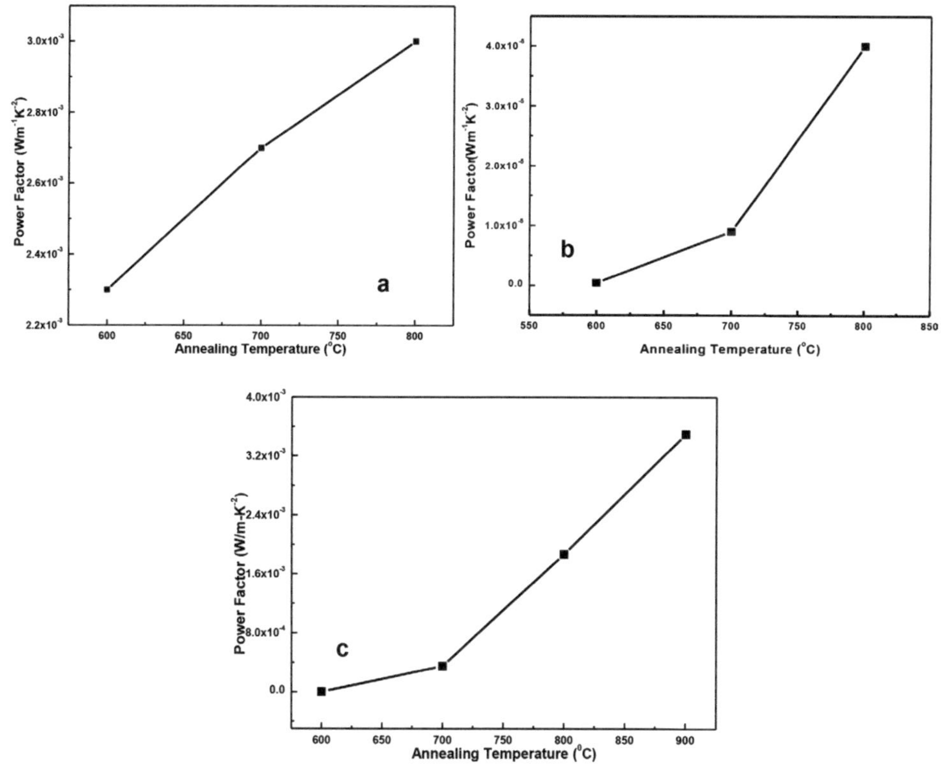

Figure 5.
(a–c) Effect of annealing temperature on the power factor of CuInO, CuAlO, and ZnGeO thin films grown by thermal evaporation, respectively.

where S is Seebeck coefficient and α is electrical conductivity. The power factor is enhanced significantly with increasing annealing and measurement temperature because both Seebeck coefficient and electrical conductivity increases.

4. Conclusion

This chapter described the effect of annealing temperature on the thermoelectric properties of oxide semiconductors. All samples were grown by thermal evaporation technique using tube furnace under vacuum using similar growth conditions. After growth, oxide semiconductors were annealed in oxygen environment at various temperatures. The reported results have suggested that Zn_2GeO_4 has good potential to be used as thermoelectric material because it has the highest value of Seebeck coefficient and power factor.

Author details

M. Ajaz Un Nabi[1], Adnan Ali[1], M. Imran Arshad[1], Sofia Akber[1], A. Manzoor[1], M. Sharif[1], Kashif Javaid[1], Nasir Amin[1], Nadeem Sabir[1], Khaalid Mahmood[1*], S. Ikram[1], Sajad Hussain[2] and H. Naeem Ur Rehman[3]

1 Department of Physics, Government College University Faisalabad, Pakistan

2 Department of Physics, Division of Science and Technology, University of Education, Pakistan

3 Department of Physics, Khawaja Fareed University of Engineering and Information Technology, Rahim Yar Khan, Pakistan

*Address all correspondence to: khalid_mahmood856@yahoo.com

References

[1] Alam H, Ramakrishna S. Nano Energy. 2013;**2**:290

[2] Elsheikh MH, Shnawah DA, Sabri MFM, Said SBM, Hassan MH, Bashir MBA, et al. Renewable and Sustainable Energy Reviews. 2014;**30**:337-355

[3] Huang M-J, Chou P-K, Lin M-C. Sensors and Actuators A: Physical. 2006;**126**:122-128

[4] Afshar O, Saidur R, Hasanuzzaman M, Jameel M. Renewable and Sustainable Energy Reviews. 2012;**16**:5639-5648

[5] Zhao D, Geng H, Teng X. Journal of Alloys and Compounds. 2012;**517**:198-203

[6] Zhu T, Li J. Progress in Materials Science. 2010;**55**:710-757

[7] Bae N-H, Han S, Lee KE, Kim B, Kim S-T. Current Applied Physics. 2011;**11**:S40-S44

[8] Xu Z, Lan Z, Zhu G, Sun K, Yu Z. Journal of Alloys and Compounds. 2012;**538**:11-15

[9] Al-Merbati AS, Yilbas BS, Sahin AZ. Applied Thermal Engineering. 2013;**50**:683-692

[10] Asenath-Smith E, Misture ST, Edwards DD. Journal of Solid State Chemistry. 2011;**184**:2167-2177

[11] Mary SA, Nair BG, Naduvath J, Okram GS, Remillard SK, Sreenivasan PV, et al. Journal of Alloys and Compounds. 2014;**600**:159

[12] Imani R, Qiu Z, Younesi R, Pazoki M, Fernandes DLA, Mitev PD, et al. Nano Energy. 2018;**49**:40-50

[13] Lu Y, Nozue T, Feng N, Sagara K, Yoshida H, Jin Y. Journal of Alloys and Compounds. 2015;**650**:558-563

[14] Jumidali MM, Sulieman KM, Hashim MR. Applied Surface Science. 2011;**257**:4890-4895

[15] Abdullah QN, Ahmed AR, Ali AM, Yam FK, Hassan Z, Bououdina M, et al. Superlattices and Microstructures. 2018;**117**:92-104

[16] Memon SS, Nafady A, Solangi AR, Al-Enizi AM, Shah MR, Sherazi ST, et al. Sensors and Actuators B: Chemical. 2018;**259**:1006-1012

[17] Kaur A, Gupta G, Ibhadon AO, Salunke DB, Sinha ASK, Kansal SK. Journal of Environmental Chemical Engineering. 2018;**6**:3621-3630

Section 2

Metastable Materials

Chapter 5

Insight into Bulk Metallic Glass Technology Development Trajectory: Mapping from Patent Information Analysis

Chih-Yuan Chen

Abstract

Bulk metallic glasses (BMGs) having a completely amorphous structure possess many attractive properties, and several groups from academia and industry have conducted research to expand their application in the market. Although many efforts have focused on investigating scientific issues related to the mechanical and chemical properties of these amorphous alloys, very few studies have assessed the development trends of these amorphous materials, especially from the viewpoint of market application and R&D directions. Therefore, in this chapter, the development trajectory of BMG materials is summarized based on data extracted from patent bibliometric information. These data were used because the information on patent documents obtained from a commercial patent database, World Intellectual Property Service (WIPS 2.0), can provide the most comprehensive information on valuable R&D activities and market issues. The results summarize advances in technology based on various alloy categories and processing routes. Furthermore, the research interests are also analyzed according to different countries, companies, and research institutions. The patent information provided in this chapter can provide a clear direction to assist metallurgist/metallurgy engineers in further technology development forecasting and R&D plan management.

Keywords: metastable material, bulk glass metal, bibliometrics, patent analysis, technological forecasting

1. Introduction

Discovered by Klement et al. in the early 1960s, amorphous metallic glasses have attracted much attention for several decades due to their outstanding properties, such as excellent mechanical properties, good corrosion resistance, and unique physical and chemical characteristics. These metallic materials are suitable for application as a new class of advanced materials [1–3]. However, their tiny size, a result of their limited glass-forming ability (GFA), makes it difficult to use BMG materials in industry.

Therefore, over the past decades, three main directions have been followed in the development of higher-quality BMG materials with better properties. These directions are as follows: (1) new compositions of metallic alloys [4–6], (2) novel processing routes [7–11], and (3) potential application fields [12–15]. In efforts to achieve new bulk metallic glasses (BMG) with high glass-forming ability (GFA), many studies have focused on establishing a relationship between the GFA and the chemical compositions of metallic materials. For example, in the Al-based metallic system, both types of metal elements, transition metal elements, and rare earth metal elements can increase the GFA of Al-based alloys [15, 16]. On the other hand, various processing routes, such as melt spinning [7], magnetron sputtering [8], pulsed laser quenching [9], and liquid splat-quenching [10] have also been developed to overcome the crucial constraint on the size and geometry of metallic glass samples. For instance, a high-throughput strategy, named the combinatorial approach via co-sputtering, has been developed for producing and characterizing substantial compositional libraries at the same time [11]. In addition, several studies have also focused on discovering potential fields of application, such as structural materials [3], hydrogen storage materials [12], soft magnetic materials [13], and biomaterials [14].

Although several research articles concerning metallic glass materials have been published, almost no studies have conducted patent analyses of metallic glass materials, to the author's best knowledge. Patent information is useful because it contains valuable research results for the researcher, business planner, R&D manager, and policymaker [17–20]. The reason is that a patent application is a costly and time-exhausting process; the willingness of the applicant to invest time, money, and effort in the process generally indicates that the patent can provide commercial benefits and technical contributions. Therefore, as pointed out by Daniel Gredel et al., patent documentation is the most comprehensive of all research resources. Nearly 70% of the technical information contained in these documents is not available in any other type of information source, and it can be used for detailed analysis [21]. For instance, patent data can be used to analyze competitors, track the evolution of technology, master crucial technologies, and identify the trends and conditions of patent development in different markets [22].

In the present research, patent data were analyzed to explore the technological development of metallic glass materials. The variations in numerous patents and assignees, technology life cycle, and categories of patents for metallic materials were studied. Furthermore, the top ten patent assignees and the trends of their patents filed, patent families, and patent citations were analyzed. The top five families and five most-cited patents are also explored in the present study.

The article is structured as follows. Section 2 presents the study methodology and the details of the information analysis. Section 3 presents an analysis of amorphous alloys patenting activity and possible explanations for the data. In Section 4, the final section, the implications and conclusions are presented.

2. Methodology

2.1 Data extraction procedure

In accordance with the suggestion of the WIPO Guide to Using Patent Information, the bibliographic records of patent documents were retrieved from a commercial database based on keywords and IPC codes related to amorphous and metallic glass alloys [23]. In the present study, we employed patent data to survey the development trends of amorphous alloy technology, as well as their technological impacts on the metallurgy industry. The patent information was gathered and analyzed by the following steps: (1) patent data retrieval, (2) patent data mining, and (3) patent data analysis.

To find the patent information, we modified the search formula as follows: [("amorphous alloy*" OR "metal*glass*" OR "glass*alloy*")] (step 1). Furthermore, the IPC search expression, namely [C*], was also included in the overall retrieval process. It should be emphasized that we performed keyword searches for the above formula appearing only in an independent claim to ensure that accurate patent information was found. Our intention was to exclude any information not involving the amorphous material industry. Patent data mining techniques were used in commercial software (WIPS 2.0) to quantify the patent information and analyze patent trends (step 2). In the final step, all the collected patent information was developed into various indicators and presented in tables or graphs plotted in commercial Microsoft software (step 3). Moreover, to avoid incorrect explanations of the trends, patents filed in 2016 and 2017 were not included in the present study, since most patents are not available before publication (i.e., for 18 months after the patent is filed). Therefore, the total number of patents that could be analyzed was 2857.

The commercial patent database WIPS 2.0 was selected for the overall search process because it includes full-text patents from 11 patent offices and abstract and bibliographic information from 75 countries. This commercial database also provides other advantages not included in other databases. For example, full-text translations of patents from Asian countries, such as China, Japan, and Korea, are provided, which facilitates complete search and analysis.

2.2 Detail of data analysis

The analyzed indicators in the present study mainly included the progress of patent documents, percentages of various categories of amorphous alloys in the development of metallic alloy technology, patent activity in different countries, technology life cycle analysis, evolution of patents filed by different assignees and countries, analysis of top ten patent assignees, analysis of technological development of the top five patent families, and five most-cited patents. It should be emphasized that the numerical indicators analyzed here were based on suggestions from the field of quantitative research in science and technology and comments from experts in the field of amorphous materials [7, 12, 15].

All evolution indicators were plotted in periods of 5 years from 1971 to 2015. Furthermore, assignees' names were also unified carefully to avoid incorrect interpretation during patent-filing trends in different countries and analysis of the top 10 assignees. To evaluate the evolution of the amorphous alloys, all IPC codes were collected and analyzed as follows: Fe as the major constituent element (C22C 45/02), Ni or Co as the major constituent element (C22C 45/04), Be as the major constituent element (C 45/06), Al as the major constituent element (C22C 45/08), and Mo, W, Nb, Ta, Ti, or Zr as the major constituent element (C22C 45/10). We further named these classifications as follows: Fe-based alloys, Ni- or Co-based alloys, Al-based alloys, and Ti- or Zr-based alloys. However, the Be-based alloys were not included in the present analysis due to their rarity (only 8 patents found).

3. Results and discussion

3.1 Amorphous alloy development: evolution of patent application

The evolution of the number of patents and assignees in every decade since the 1970s is shown in **Figures 1** and **2**, respectively. Both figures show obvious increments in the numbers of both patents and assignees after 1990. From subsequent analysis, it was found that the first apparent growth in the number of patent

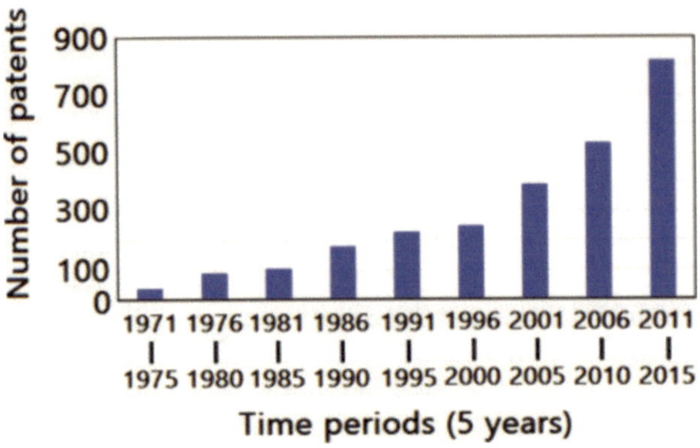

Figure 1.
Evolution of the number of patents.

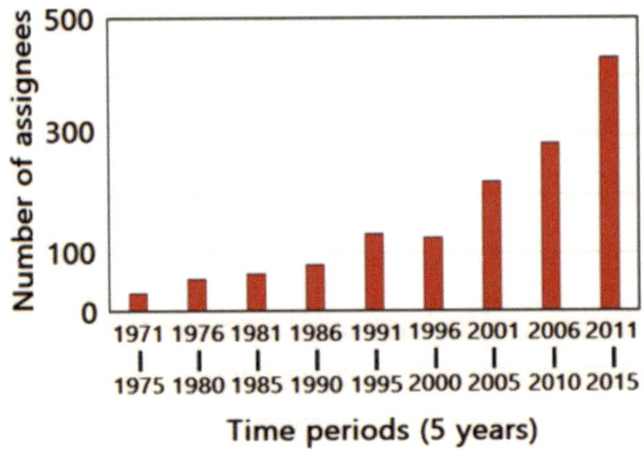

Figure 2.
Evolution of the number of assignees.

applications/assignees was indicated by an increase in the number of patents filed in Japan in 1990. The second surge in patent applications/assignees occurred in 2000 as a consequence of patent filings from China.

The trend of patenting different categories of amorphous alloys is shown in **Figure 3**. Fe-based amorphous alloys had the most active patenting activity in the

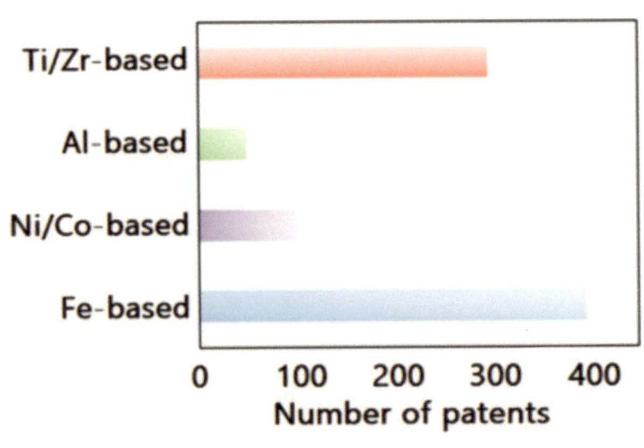

Figure 3.
Patenting activity for different amorphous alloys.

period of investigation. These Fe-based patents were filed by mostly Japanese corporations and research institutes mainly due to their magnetic properties, which can be exploited in the technology fields of electric/magnetic devices, energy storage devices, and some semiconductor devices [24, 25]. On the other hand, more than Ni-/Co-based and Al-based amorphous alloys, Ti-based and/or Zr-based amorphous alloys occupied the second role of patenting activity because these amorphous alloys possess superior mechanical properties and biocompatibility and can be widely used in the biomedical industry, even though applications in electronic and energy conversion devices were also widely found [26, 27].

Figure 4 shows the technology life cycle curve for amorphous alloys in the present study. Grasping the technology life cycle curve is critical for researchers and R&D managers in assessing further trends in technological development. As pointed out by Trappey et al., a typical technology life cycle curve can be divided into four stages: (1) introductory, (2) growth, (3) maturity, and (4) decline [28]. In the introductory stage, the number of patents and assignees are extremely low because very few corporations have invested in the field. In the following years, more and more assignees become involved in the prospective technology field due to a reduction of uncertainty in the market and technology outlook, which also leads to a gradual increase in patent applications as the life cycle moves into the growth stage. On the other hand, if the number of patents and the number of assignees begin to decrease, the stage is classified as technological maturity. Only a few corporations are willing to invest in such a technology.

Therefore, in the present study, the rapid increase in the numbers of patents and assignees after 2000 indicated that the technology had entered the growth stage. The average numbers of patents and assignees, respectively, increased from 47 and 52 in 2000 to 157 and 86 in 2015. The characteristic of this growth stage is the absence of technical problems and market uncertainty, which leads to more companies becoming involved in developing related products for the market. This stage is a possible explanation for the surge in patenting in China.

3.2 Amorphous alloy development: analysis by country and assignee nationality

Figure 5 shows the number of patent application in various countries and their evolutionary trends. As shown in the figure, China, Japan, the United States, Korea, and Europe were the top five patent-filing countries/regions, indicating their potential market attraction. For example, the total number of patent applications in China was

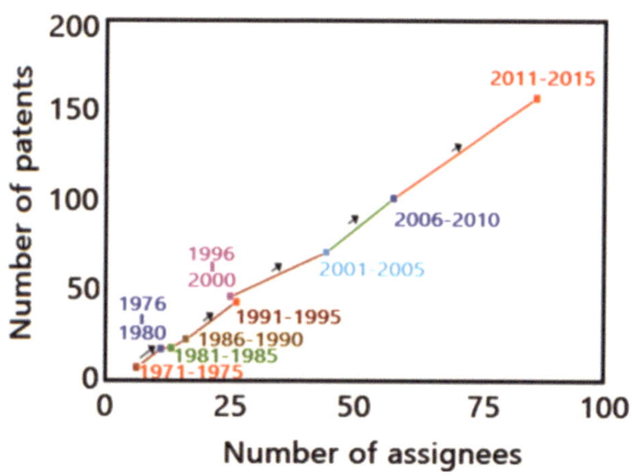

Figure 4.
Technology life cycle.

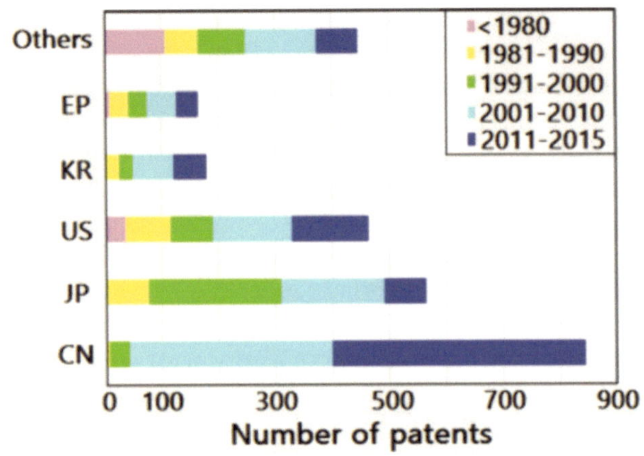

Figure 5.
Number of patents and its evolution by country.

844, which implies the commercial value of bulk metallic glasses in the Chinese market. Another interesting event in the same figure is the R&D investment transfer from the United States and Japan to China. Discovered in the United States and further exploited in Japan, the number of patents related to metallic glass alloys in Japan in the last three decades gradually decreased from 233 (1991–2000) to 181 (2001–2010) and then to 73 (2011–2015). In contrast, unlike those in other countries, the patents filed in China increased by a wide margin, from 35 (1991–2000) to 358 (2001–2010) and then 444 (2011–2015). The above surge in patent filing in China, often called China's patent boom, occurred not only in amorphous alloy technology but also in other technological areas. As pointed out by several research teams, a decrease in the cost of filing a patent and the initiation of a subsidy program for patent applications were two major causes of the explosion in patent applications in China [29, 30].

Like that in China, patenting behavior in the United States shows a similar trend of progressive increases in the numbers of patents filed over the past three decades: from 72 (1991–2000) to 140 (2001–2010) and finally to 135 (2011–2015). In Korea and Europe, however, the number of patents filed has declined since 2011. The explanation of phenomenon is still unclear and will require further study.

3.3 Analysis by top ten patent assignees

Analysis of the patent assignees from the original patent data pool can help researchers to understand the technological development strategies and product development trajectories of large companies. Therefore, the top ten patent assignees with a focus on the development of amorphous alloys are summarized in **Table 1**. The top ten patent assignees were mainly from Japan (7), China (2), and the United States (1). Moreover, the assignee from the United States was an academic institution, whereas those from Japan and China were mostly corporations and research institutions.

The number of patent families and average number of countries where patents were filed are also shown in **Table 1**. The variation in patent families was related to the total number of patents; that is, the patent families increased monotonically in conjunction with the number of patent applications. The top assignee, California Institute of Technology (CIT), was associated with 633 patent families, which is obviously higher than those of the other assignees. The reason for this large difference in the number of patent families could be attributed to the fact that CIT filed its patents in many countries (an average of 7). In contrast, Chinese assignees generally filed patents only in their home country, which explains the similarity between the number of patents and patent families.

Assignee (nationality)	Total number of patents	Patent families	Average cited times	Average Number of application countries
California Inst. Techn. (US)	96	633	16.0	7
Hitachi Metals LTD (JP)	76	376	4.7	5
BYD Company LTD (CN)	68	134	2.1	2
Univ. Tohoku (JP)	67	221	2.0	3
Nippon Steel Corp. (JP)	67	270	3.5	4
YKK Corp. (JP)	62	252	4.1	4
ALPS Electric Co. LTD (JP)	59	192	3.9	3
Japan Science & Tech Corp. (JP)	45	103	6.9	2
Inst. Metal Res. Chinese Acad. Sc. (CN)	35	48	1.9	1
JFE Steel KK (JP)	29	93	3.5	3

Table 1.
Top ten patent assignees.

Table 1 also presents the average number of citations of each assignee's patents; this number can be used to assess the quality of a patent [31, 32]. A weak relationship was found to exist between the average number of citations and the total number of patents/patent families. Therefore, a high number of citations indicate that more related inventions followed, which usually implies a higher economic value of the patent. In addition, patents filed by CIT got the most attention. They had the highest number of citations (16.0), which was considerably higher than those of other assignees. In contrast, patents owned by China were rarely cited, which could be associated with their short filing periods (**Table 2**).

3.4 Technological development strategy analysis: analysis of the top five patent families

The analysis of the top five patent families is shown in **Table 3**. The first patent family, US2009236017A1 [33], proposes an apparatus and method comprising

Assignee (number of active years)	1981 ~1990	1991 ~2000	2001 ~2010	2011 ~2015	Total
California Inst. Techn. (20)	-	18	31	43	92
Hitachi Metals LTD. (27)	3	16	32	18	69
BYD Company Limited (9)	-	-	46	22	68
UNIV. Tohoku (12)	-	-	51	14	65
Nippon Steel CORP. (22)	5	24	35	3	67
YKK CORP (15)	3	51	8	-	62
ALPS Electric Co. LTD. (15)	11	45	3	-	59
Japan Science & Tech. Corp. (13)	-	16	29	-	45
Inst. Metal Res. Chinese Acad. Sc. (14)	-	1	14	12	27
JFE Steel KK (10)	-	23	-	5	28

Table 2.
Trend of the patent applications for top ten patent assignees.

Patent number [reference]/ assignee (nationality)	Title	Patent Families
US2009236017A1 [33]/Cal. Inst. Tech. (US)	Forming of metallic glass by rapid capacitor discharge	57
US7323071B1 [34]/Bechtel BWXT Idaho, LLC. (US)	Method for forming a hardened surface on a substrate	33
US5628840 [35]/Allied Signal INC. (US)	Metallic glass alloys for mechanically resonant surveillance systems	29
US6183568B1 [36]/Fuji Photo Film Co. LTD (JP)	Method for preparing a magnetic thin film	25
US8529712B2 [37]/Cal. Inst. Tech. (US)	Tough iron-based bulk metallic glass alloys	21

Table 3.
Top five patent families.

uniform heating, rheological softening, and thermoplastic forming of metallic glasses rapidly into a net shape using a rapid capacitor discharge forming (RCDF) tool. The RCDF method utilizes a discharge of electrical energy stored in a capacitor to uniformly and rapidly heat a sample or charge of metallic glass alloy to a predetermined "process temperature" between the glass transition temperature of the amorphous material and the equilibrium melting point of the alloy on a time scale of several milliseconds or less. Once the sample is uniformly heated such that the entire sample block has a sufficiently low process viscosity, it may be shaped into high-quality amorphous bulk articles by any number of techniques, such as injection molding, dynamic forging, stamp forging, and blow molding, in a time frame of less than 1 s. The second patent family, US7323071B1 [34], discloses a metallic glass coating formed over a metallic substrate. After the formation of the coating, at least a portion of the metallic glass can be converted into a crystalline material having a nanocrystalline grain size. The third patent family, US5628840 [35], relates to a glassy metal alloy consisting essentially of the formula $Fe_aCo_bNi_cM_dB_eSi_fC_g$, where "M" is at least one member selected from the group consisting of molybdenum, chromium, and manganese. The notations "a–g" are in atom percent, "a" ranges from about 30 to about 45, "b" ranges from about 4 to about 40, "c" ranges from about 5 to about 45, "d" ranges from about 0 to about 3, "e" ranges from about 10 to about 25, "f" ranges from about 0 to about 15, and "g" ranges from about 0 to about 2. The alloy can be cast by rapid solidification into a ribbon or otherwise formed into a marker that is especially suited for use in magneto-mechanically actuated article surveillance systems. Advantageously, the marker is characterized by relatively linear magnetization response in the frequency regime wherein harmonic marker systems operate magnetically. Voltage amplitudes detected for the marker are high, and interference between surveillance systems based on mechanical resonance and harmonic re-radiance is virtually eliminated. The fourth patent family, US6183568B1 [36], proposes a soft magnetic thin microcrystalline film of $Fe_aB_bN_c$ (at %) where B is at least one of Zr, Hf, Ti, Nb, Ta, V, Mo, and W, and $0 < b \leqq 20$ and $0 < c \leqq 22$, except for the range of $b \leqq 7.5$ and $c \leqq 5$, show low coercivity Hc of 80–400 Am^{-1} (1–5 Oe), which is stable upon heating at elevated temperature for glass bonding. This film is produced by crystallizing an amorphous alloy film of a similar composition at 350–650°C to a crystal grain size of up to 30 nm to provide uniaxial anisotropy and increased magnetic permeability at the higher frequency. It can also provide low magnetostriction of around $\lambda s = 0$. The composite magnetic head is made using this thin film. A diffusion-preventive SiO2 layer disposed between the ferrite core, and this thin film in the magnetic head prevents an interdiffusion layer and suppresses beats in the output signal. The fifth patent family, US8529712B2 [37], relates to an iron-based bulk metallic glass alloy and more particularly to a family of the iron-based phosphor-containing bulk metallic glass alloys exhibiting low shear moduli. The independent claim specifies an Fe-based metallic glass composition comprising at least Fe, Mo, P, C, and B, where Fe comprises an atomic percent of at least 60, Mo comprises an atomic percent of 2–8, P comprises an atomic percent of 5–17.5, C comprises an atomic percent of 3–6.5, and B comprises an atomic percent of 1–3.5, wherein the composition has a shear modulus (G) of less than 60 GPa and a glass transition temperature (T_g) of less than 440° C, and the composition is capable of forming a bulk amorphous object having a critical thickness of at least 2 mm.

3.5 Technological exploitation analysis: five most-cited patents

In most scientific publications, patents are rarely cited in SCI papers. For example, only about 1.5% of US patents are cited in SCI journals [38]. Similarly, in most technology fields, most of the prior art cited within patents are also patent documents, not scientific papers, which could be a sign of the few connections between academia and industry [38, 39]. However, the technological value of patents can provide important information to subsequent researchers and thus is worthwhile to refer to, especially in the case of patents with a high number of citations. Therefore, like a scientific paper, a high number of citations represent the high technological value of a patent, which might indicate that high commercial profit can be expected. In the present study, the five most-cited patents were extracted from the original patent pool (**Table 4**), and their technological contents are reviewed as follows.

The most-cited patent in **Table 4**, US5288344, is about beryllium-bearing amorphous metallic alloy formed with a low cooling rate [40]. In this patent, the proposed technology suggests an alloy containing beryllium in the range of 5–52 atomic percent and at least one early transition metal in the range of 30–75 atomic percent and at least one late transition metal in the range of 2–52 atomic percent. A preferred group of metallic glass alloys has the formula $(Zr_{1-x}Ti_x)_a(Cu_{1-y}Ni_y)_bBe_c$. A preferred embodiment is a class of alloys which form metallic glass upon cooling below the glass transition temperature at a rate of less than 10^3 K/s, which is far below the normal cooling rate, 10^4 to 10^6 K/s. The second most-cited patent, US5735975, describes at least quinary alloys that form metallic glass upon cooling below the glass transition temperature at a rate of less than 10^3 K/s [41]. Such alloys comprise zirconium

Patent number [reference]/ assignee (nationality)	Title	Times Cited/ (Patent family)
US5288344 [40] / Cal. Inst. Tech. (US)	Beryllium bearing amorphous metallic alloys by low cooling rates	290 / (12)
US5735975 [41] / Cal. Inst. Tech. (US)	Quinary Metallic glass alloys	220 / (7)
US5368659 [42] / Cal. Inst. Tech. (US)	Method of forming beryllium bearing metallic glass	204 / (12)
US5618359 [43] / Cal. Inst. Tech. (US)	Metallic glass alloys of Zr, Ti, Cu, and Ni	169 / (7)
US6325868 [44] / Univ. Yonsei Seoul (KR)	Nickel-based amorphous alloy compositions	157 / (4)

Table 4.
Top five patent families.

and/or hafnium in the range of 45–65 atomic percent, titanium and/or niobium in the range of 4–7.5 atomic percent, and aluminum and/or zinc in the range of 5–15 atomic percent. The balance of the alloy composition comprises copper, iron, cobalt, and nickel. The composition is constrained such that the atomic percent of iron is less than 10%. Furthermore, the ratio of copper to nickel and/or cobalt is in the range of 1:2 to 2:1. Therefore, the alloy glass can be formed at a reduced critical cooling rate without any beryllium addition. The third most-cited patent, US5368659, discloses an invention similar to the above most-cited patent; the alloy forms metallic glass containing beryllium in a narrower range of 2–47 atomic percent, at least one early transition metal in the range of 30–75 atomic percent, and at least one late transition metal in the range of 5–62 atomic percent [42]. Furthermore, the critical cooling rate to achieve the amorphous structure can be reduced to 1–100 K/s or lower. Patent US6325868, the fifth most-cited patent, discloses a nickel-based amorphous alloy composition, particularly a quaternary nickel-based amorphous alloy containing nickel, zirconium, and titanium as the main constituent elements and additive Si or P [44]. The quaternary nickel-zirconium-titanium-phosphorus alloy compositions comprise nickel in the range of 50–62 atomic percent, zirconium and titanium in the range of 33–46 atomic percent, and phosphorus in the range of 3–8 atomic percent, represented by the general formula $Ni_d(Zr_{1-y}Ti_y)_eP_f$. The nickel-based amorphous alloy compositions have a superior amorphous phase-forming ability, and bulk amorphous alloy having a thickness of 1 mm can be produced by general casting methods.

4. Conclusion

This study focuses on the analysis of patent data to explore the technological developments of metallic glass materials. The primary findings of this study can be summarized as follows:

1. Two obvious increments in patent applications occurred in 1990–2000 and 2000–2015. The former primarily arose from the increased number of patent applications filed in Japan. The latter is mainly attributed to patent applications filed in China.

2. Patents related to iron-based alloys are the main category of metallic glass materials for which patents were applied. The reason is that iron-based metallic materials can be used in several applications, especially in the soft magnetic fields.

3. According to the analysis of the present study, metallic glass materials are in the growth stage of the technology life cycle, which implies that increasing amounts of resources will be invested in the metallic glass field for the development of commercial products.

4. All of the top ten assignees were from the United States, China, and Japan. The US assignees were from an academic institution, whereas most of their Japanese and Chinese counterparts were from commercial businesses or research institutions.

5. Patents field by the California Institute of Technology were cited the most frequently, with an average of 16.0 citations, implying their high technological value.

Acknowledgements

This work was performed with financial support from the Ministry of Science and Technology of the Republic of China, Taiwan, under Contract MOST 106-2218-E-027-011-MY3.

Author details

Chih-Yuan Chen
Graduate Institute of Intellectual Property, National Taipei University of Technology, Taipei, Taiwan

*Address all correspondence to: chen6563@gmail.com

References

[1] Klement W, Willens RH, Duwez P. Non-crystalline structure in solidified gold-silicon alloys. Nature. 1960;**187**:869-870. DOI: 10.1038/187869b0

[2] Wang WH. Bulk metallic glasses with functional physical properties. Advanced Materials. 2009;**21**:4524-4544. DOI: 10.1002/adma.200901053

[3] Kruzic JJ. Bulk metallic glasses as structural materials: A review. Advanced Engineering Materials. 2016;**18**:1308-1331. DOI: 10.1002/adem.201600066

[4] Ansariniya M, Seifoddini A, Hasani S. $(Fe_{0.9}Ni_{0.1})_{77}Mo_5P_9C_{7.5}B_{1.5}$ bulk metallic glass matrix composite produced by partial crystallization: The non-isothermal kinetic analysis. Journal of Alloys and Compounds. 2018;**763**:606-612. DOI: 10.1016/j.jallcom.2018.05.360

[5] Fan C, Liu CT, Chen G, Chen G, Chen D, Yang X, et al. Influence of the molten quenching temperature on the thermal physical behavior of quenched Zr-based metallic glasses. Intermetallics. 2013;**38**:19-22. DOI: 10.1016/j.intermet.2013.02.004

[6] Zhou M, Zhou J, Wei J, Yang M, Ma L. Enhanced glass-forming ability and mechanical properties of $Zr_{65}Cu_{17.5}Al_{7.5}Ni_{10}$ metallic glass by adding Fe. Journal of Non-Crystalline Solids. 2017;**455**:1-5. DOI: 10.1016/j.jnoncrysol.2016.05.004

[7] Saage G, Roth S, Eckert J, Schultz L. Low magnetostriction crystalline ribbons prepared by melt–spinning and reactive annealing. Materials Science and Engineering A. 2004;**375-377**:1125-1128. DOI: 10.1016/j.msea.2003.10.284

[8] Nayebossadri S, Smith D, Speight J, Book D. Amorphous Zr-based thin films fabricated by magnetron sputtering for potential application in hydrogen purification. Journal of Alloys and Compounds. 2015;**645**:S56-S60. DOI: 10.1016/j.jallcom.2015.01.230

[9] Jaffari GH, Rumaiz AK, Ni C, Yassitepe E, Bah M, Shah SI. Observation of metastable phase separation and amorphous phase in $Fe_{67}Co_{33}$ alloy thin films synthesized by pulsed laser depositions. Current Applied Physics. 2015;**15**:717-721. DOI: 10.1016/j.cap.2015.03.001

[10] Poon SJ, Dowling TE. Superconductivity of the allotropic forms of zirconium-based alloys obtained by liquid quenching on hot substrates. Solid State Communications. 1984;**50**:189-191. DOI: 10.1016/0038-1098(84)90937-2

[11] Coddet P, Sanchette F, Rousset JC, Rapaud O, Coddet C. On the elastic modulus and hardness of co-sputtered Zr–Cu–(N) thin metal glass films. Surface and Coating Technology. 2012;**206**:3567-3571. DOI: 10.1016/j.surfcoat.2012.02.036

[12] Zhang B, Lv Y, Yuan J, Wu Y. Effects of microstructure on the hydrogen storage properties of the melt-spun Mg-5Ni-3La (at. %) alloys. Journal of Alloys and Compounds. 2017;**702**:126-131. DOI: 10.1016/j.jallcom.2017.01.221

[13] Wei R, Sun H, Chen C, Tao J, Li F. Formation of soft magnetic high entropy amorphous alloys composites containing in situ solid solution phase. Journal of Magnetism and Magnetic Materials. 2018;**449**:63-67. DOI: 10.1016/j.jmmm.2017.09.065

[14] Li HF, Zheng YF. Recent advances in bulk metallic glasses for biomedical applications. Acta Biomaterialia. 2016;**36**:1-20. DOI: 10.1016/j.actbio.2016.03.047

[15] Chen C-Y, Hsu W-S, Chen S-F. Role of Sc on the glass forming ability and mechanical properties of Al-Y-Ni-Sc bulk metallic glass produced with different cooling rates. Materials Science and Engineering A. 2018;**725**:119-126. DOI: 10.1016/j.msea.2018.04.003

[16] Chen S-F, Chen C-Y, Lin C-H. Insight on the glass-forming ability of Al–Y–Ni–Ce bulk metallic glass. Journal of Alloys and Compounds. 2015;**637**:418-425. DOI: 10.1016/j.jallcom.2015.02.217

[17] Yang YY, Klose T, Lippy J, Barcelon-Yang CS, Zhang L. Leveraging text analytics in patent analysis to empower business decisions—A competitive differentiation of kinase assay technology platforms by I2E text mining software. World Patent Information. 2014;**39**:24-34. DOI: 10.1016/j.wpi.2014.09.002

[18] Kang B, Motohashi K. The role of essential patents as knowledge input for future R&D. World Patent Information. 2014;**38**:33-41. DOI: 10.1016/j.wpi.2014.05.001

[19] Lee S, Yoon B, Lee C, Park J. Business planning based on technological capabilities: Patent analysis for technology-driven roadmapping. Technological Forecasting and Social Change. 2009;**76**:769-786. DOI: 10.1016/j.techfore.2009.01.003

[20] Schleich J, Walz R, Ragwitz M. Effects of policies on patenting in wind-power technologies. Energy Policy. 2017;**108**:684-695. DOI: 10.1016/j.enpol.2017.06.043

[21] Gredel D, Kramer M, Bend B. Patent-based investment funds as innovation intermediaries for SMEs: In-depth analysis of reciprocal interactions, motives and fallacies. Technovation. 2012;**32**:536-549. DOI: 10.1016/j.technovation.2011.09.008

[22] Mogee ME. Using patent data for technology analysis and planning. Research-Technology Management. 1991;**34**:43-49. DOI: 10.1080/08956308.1991.11670755

[23] WIPO. WIPO Guide to Using Patent Information. Geneva: WIPO; 2012

[24] Fujikura M, Yamada T, Sato T. Improvement of magnetic properties by addition of tin to amorphous Fe-Si-B alloys with high iron contents. Materials Science and Engineering A. 1994;**181-182**:1351-1354. DOI: 10.1016/0921-5093(94)90861-3

[25] Kronmüller H, Ignatchenko VA, Forkl A. Law of approach to magnetic saturation in inhomogeneous ferri- and antiferromagnets and in amorphous iron-rich Fe-Zr alloys. Journal of Magnetism and Magnetic Materials. 1993;**134**:68-74. DOI: 10.1016/0304-8853(94)90074-4

[26] Liu L, Chan KC, Yu Y, Chen Q. Bio-activation of Ni-free Zr-based bulk metallic glass by surface modification. Intermetallics. 2010;**18**:1978-1982. DOI: 10.1016/j.intermet.2010.02.039

[27] Lai JJ, Lin YS, Chang CH, Wei TY, Huang JC, Liao ZX, et al. Promising Ta-Ti-Zr-Si metallic glass coating without cytotoxic elements for bio-implant applications. Applied Surface Science. 2018;**427**:485-495. DOI: 10.1016/j.apsusc.2017.08.065

[28] Trappey CV, Trappey AJ, Wu CY. Clustering patents using non-exhaustive overlaps. Journal of Systems Science and Systems Engineering. 2010;**19**:162-181. DOI: 10.1007/s11518-010-5134-x

[29] Hu AGZ, Zhang P, Zhao L. China as number one? Evidence from China's most recent patenting surge. Journal of Development Economics. 2017;**124**:107-119. DOI: 10.1016/j.jdeveco.2016.09.004

[30] Dang J, Motohashi K. Patent statistics: A good indicator for innovation

in China? Patent subsidy program impacts on patent quality. China Economic Review. 2015;**35**:137-155. DOI: 10.1016/j.chieco.2015.03.012

[31] Thompson MJ. Measuring patent quality: A claim and search report approach. World Patent Information. 2016;**45**:47-54. DOI: 10.1016/j.wpi.2016.03.003

[32] Sterzi V. Patent quality and ownership: An analysis of UK faculty patenting. Research Policy. 2013;**42**: 564-576. DOI: 10.1016/j.respol.2012.07.010

[33] Johnson WL, Demetriou MD, Kim CP, Schramm JP. Forming of metallic glass by rapid capacitor discharge. US Patent: US2009236017A1; 2009

[34] Branagan DJ. Method for forming a hardened surface on a substrate. US Patent: US7323071B1; 2008

[35] Hasegawa R. Metallic glass alloys for mechanically resonant marker surveillance systems. US Patent: US5628840; 1997

[36] Nakanishi K, Shimizu O, Yoshida S, Katayama M, Isomura T. Method for preparing a magnetic thin film. US Patent: US6183568B1; 2001

[37] Demetriou MD, Johnson WL. Tough iron-based bulk metallic glass alloys. US Patent: US8529712B2; 2013

[38] Glänzel W, Meyer M. Patents cited in the scientific literature: An exploratory study of 'reverse' citation relations. Scientometrics. 2003;**58**: 415-428. DOI: 10.1023/A:1026248929668

[39] Meyer M. Does science push technology? Patents citing scientific literature. Research Policy. 2000;**29**:409-434. DOI: 10.1016/S0048-7333(99)00040-2

[40] Peker A, Johnson WL. Beryllium bearing amorphous metallic alloys formed by low cooling rates. US Patent: US5288344; 1994

[41] Lin X, Johnson WL. Quinary metallic glass alloys. US Patent: US5735975; 1998

[42] Peker A, Johnson WL. Method of forming berryllium bearing metallic glass. US Patent: US5368659; 1994

[43] Lin X, Peker A, Johnson WL. Metallic glass alloys of Zr, Ti, Cu and Ni. US Patent: US5618359; 1997

[44] Kim DH, Kim WT, Yi SH, Lee JK, Lee MH, Park TG, et al. Nickel-based amorphous alloy compositions. US Patent: US6325868B1; 2001

Chapter 6

Calculation of the Metastable Atom Densities in Argon and Neon Abnormal Glow Discharges

Abdelaziz Bouchikhi

Abstract

In this chapter an investigation of a DC argon and neon abnormal glow discharges with metastable atom density is presented. The values of pressure are between 133.32 and 330 Pa, and the voltage range is from 250 to 400 V in the case of argon gas. In the case of neon gas, the pressure has the value of 399.92 Pa (3 Torr) and the voltage ranges from 300 to 500 V. In the frameworks, an analysis of abnormal glow discharge characteristics is carried out in the case of input data taken from the Boltzmann equation in multi term approximation (BMA), and in the case of input data obtained from BOLISG+ code. As conclusion of these differences of input data in the same gas the output results are different and it appears in the cathodic region. The spatiotemporal distributions of electron and ion densities, the potential and electric field, the mean electron energy and the metastable atom density are shown. A 1D fluid model is used to solve self-consistently the first three moments of the Boltzmann's equation coupled with the Poisson's equation. Our results are validated with those obtained by both recent paper and experimental results.

Keywords: metastable atom density, abnormal glow discharge, fluid model, input data

1. Introduction

The concern of the amelioration of the plasma reactor is always a domain important in the development technology, among these fields we find glow discharge. A plasma technology in a gas mixture has been studied by several authors. Ono et al. [1] have been studied oxygen-nitrogen gas mixture glow discharge plasma by intervene many chemical reactions in their model. Khomich et al. [2] have been treated the problem of the atomic deposition in the metal surface modification by nitrogen-argon mixture glow discharge in abnormal regime. Ponduri et al. [3] have been analyzed the dissociation of CO_2 by dielectric barrier glow discharge, as a consequence of utilization of CO_2 gas a lot of kinds of species intervene in the phenomena discharge. Baadj et al. have been [4] investigated Xe-Cl_2 gas mixture for the formation of XeCl* exciplex lamp by means of zero-dimensional model. Li et al. [5] have been studied the plasma jet length in Ne, Ar, He and Kr in atmospheric pressure when the excimer molecule formatted from

IntechOpen

metastable state of the atomic gases, and they have been identified three modes versus of the gas flow rate, its about laminar, transition, and turbulent jet modes.

In this chapter, an research is through concerning the role of neon and argon metastable atoms in the discharge. Metastable atoms have been considered by several authors, both theoretically and experimentally. Metastable densities can experimentally be measured by optical absorption method. In theoretical means, a poise equation, including different production and loss terms is assembled to compute the metastable densities. Experimental measurements were executed, for example, for Neon gas in an RF glow discharge by Eckstein et al. [6], for Ar gas in a microwave boosted glow discharge by Uzelac and Leis [7], and for He gas as a function of discharge conditions by Browne and Dunn [8]. In the research of Smith et al. [9] purely relative absorption signals of Argon metastable atoms as a function of current and pressure were experimented. In the research of Strauss et al. [10] and Ferreira et al. [11] Argon metastable densities were experimented in a afterglow of a pulsed discharge and in the Grimm-type glow discharge, respectively, and some mechanisms for the structure of Argon metastable atoms were recommended. Argon metastable densities have also been experimented by Ferreira et al. [12], and by Ferreira and Ricard [13].

A comparison was complete with Argon metastable densities determined from a coupled-electron-metastable-atom model. Hardy and Sheldon [14] have examined Argon, Helium and Neon gases. A comparison between calculated and measured metastable densities was also realized by Kubota et al. [15] for Helium in a RF and DC glow discharge. Lymberopoulos and Economou [16] have established a combined fluid model for the electrons, Ar metastable atoms, and Ar ions in order to study the effect of metastable atoms in the discharge. In the works [17–22] rate constants of a number of collision processes in control of the demolition of metastable atoms were got by combining equilibrium equations with the experimented time-dependent variation of the metastable densities or by analyzing the reliance of the decay constants upon pressure. Den Hartog et al. [23, 24] have investigated Helium gas.

Last-mentioned, Fedoseev and Sukhinin [25] have investigated the influence of metastable Ar atoms on gas discharge plasma with dust particles. Shumova et al. [26] have investigated the effect of metastable Ne atoms and dust particles in a positive column of glow discharge.

The aim of this work is to present, influence of the discharge characteristic in the case of input data obtained by BOLSIG+ code, and in the case of input data taken from the Boltzmann equation in multi term approximation (BMA). We note that both these approaches are widely used. For simulations of positive column of glow discharge, Vasilyak et al. [27] used the first approach, while Sukhinin et al. [28] used the second approach. In Section 2, the mathematical model is delineated; it comprises the boundary and initial conditions as well as the numerical method. In Section 3, the results are discussed for Ar discharge. In Section 4, the test scheme is given. In Section 5, Influence of the input data got from BOLSIG+ on the argon discharge has been shown. In Section 6, the characteristics of the Ne discharge with input data taken from BOLSIG+ are presented. Finally, the conclusion of the chapter is given in Section 7.

2. Discharge modeling

Our mathematical model builds on the first three moments of the Boltzmann equation. The Continuity equations and momentum transfer equations of metastable atom, electrons and positively charged ions. The energy equation is known only for electrons, and to give the effect of the electric field on the particles charged, the Poisson equation is included in the model [29, 30].

The chemical reactions intervene in the discharge are indicated in the **Table 1**.

After that, the model in the 1D Cartesian geometry, writes by partial differential equations:

$$\frac{\partial n_e}{\partial t}+\frac{\partial \varphi_e}{\partial x}=n_e\left(n_o K_o^{io}+n_m K_m^{io}\right)+n_m n_m K^{ci}, \quad (1)$$

$$\frac{\partial n_+}{\partial t}+\frac{\partial \varphi_+}{\partial x}=n_e\left(n_o K_o^{io}+n_m K_m^{io}\right)+n_m n_m K^{ci}, \quad (2)$$

$$\frac{\partial n_m}{\partial t}+\frac{\partial \varphi_m}{\partial x}=n_e\left(n_o K_o^{m}-n_m K_m^{o}-n_m K_m^{io}\right)-2n_m n_m K^{ci}-\frac{n_m}{\tau_m}, \quad (3)$$

$$\begin{aligned}\frac{\partial \varepsilon_e n_e}{\partial t}+\frac{\partial \varphi_{\varepsilon\varepsilon}}{\partial x}&=-e\varphi_e E+\varepsilon^m n_e n_m K_m^o+\varepsilon^{ci} n_m n_m K^{ci}\\ &-n_e P^{ec}-n_e\left(\varepsilon^m n_o K_o^m+\varepsilon^{io} n_o K_o^{io}+\left(\varepsilon^{io}-\varepsilon^m\right)n_m K_m^{io}\right)\end{aligned}, \quad (4)$$

$$\frac{\partial^2 V}{\partial x^2}=-\frac{e}{\varepsilon_o}\left(n_+-n_e\right) \quad (5)$$

Here n_e, n_+, n_m, φ_m, φ_e and φ_+ are number densities, densities of transport flux of the metastable atoms, electrons and ions, respectively. K_o^m is the rate coefficient of electron-impact excitation of ground state atoms, $\varepsilon^m = 11.55$ eV is the energy loss of excited atoms, n_o is the constant background gas density, K_m^{io} is the rate coefficient of electron-impact ionization of excited atoms with the energy loss $\left(\varepsilon^{io}-\varepsilon^m\right)$, K_o^{io} is the rate coefficient of electron-impact ionization of ground state atoms and $\varepsilon^{io} = 15.76$ eV is the energy loss of ionized atoms, K_m^o is the rate coefficient of de-excitation of excited atoms by electron collisions. $E=-\partial V/\partial x$ is the electric field strength. ε_o and e are the permittivity of free space and elementary charge, respectively. $K^{ci} = 8.1\times 10^{-10}$ cm^3 s^{-1} [31] is the rate coefficient of chemo-ionization processes with the energy gain $\varepsilon^{ci}=2\varepsilon^m-\varepsilon^{io}$. $\tau_m = 1$ μs is the metastable lifetime. ε_e is the mean electron energy, $\varphi_{\varepsilon\varepsilon}$ is the electron energy flux. V is the electrostatic potential. P^{ec} is the energy loss per electron due to elastic collision of electrons with the background gas [32].

Momentum transfer equations for metastable atoms, electrons, ions and electron energy [33–35] are:

Processes	Name of processes
$Ar + e^- \rightarrow Ar^+ + 2e^-$	Ionization
$Ar + e^- \rightarrow Ar_m^* + e^-$	Excitation
$Ar_m^* + e^- \rightarrow Ar + e^-$	De-excitation
$e^- + Ar_m^* \rightarrow Ar^+ + 2e^-$	Stepwise ionization
$Ar_m^* + Ar_m^* \rightarrow Ar^+ + e^- + Ar$	Chemo-ionization

Table 1.
Kinetic scheme of processes.

$$\varphi_e = -n_e\mu_e E - \frac{\partial D_e n_e}{\partial x}, \quad (6)$$

$$\varphi_+ = n_+\mu_+ E - \frac{\partial D_+ n_+}{\partial x}, \quad (7)$$

$$\varphi_m = -D_m \frac{\partial n_m}{\partial x}, \quad (8)$$

$$\varphi_{e\varepsilon} = -n_e E\mu_{e\varepsilon} - \frac{\partial n_e D_{e\varepsilon}}{\partial x} \quad (9)$$

Here μ_e, μ_+, D_e and D_+ are the electron, ion for mobilities and diffusion coefficients, respectively. $D_{e\varepsilon}$ and $\mu_{e\varepsilon}$ are the diffusivity and mobility of electron energy transport. The ion mobility has brought from Phelps and Petrović [36] and their ion diffusivity is computed agree with Einstein's relation [37]. The coefficients for electrons [38] in argon as dependences on the mean electron energy are got as of INP Greifswald for direct evaluation with the results obtained by Becker et al. [30]. D_m is the metastable atom diffusivity someplace $n_o D_m = 1.7x10^{18}$ cm^{-1} s^{-1} [19].

2.1 Boundary and initial conditions

The discharge is affected between two parallel plate electrodes and the radius of the electrode is presumed to be higher than the electrode gap and the physical characteristic distributions are approximately uniform along the radial direction. The grounded electrode has been put at $x = 1$ cm, play the role as the anode ($V_{anode} = 0$). The powered electrode has been put at $x = 0$ cm, which initiates the model discretization, play the role as the cathode ($V_{cathode} = -V_{DC}$).

At time $t = 0$, the metastable atom, electron, and ion densities are supposed constant and equal to 10^3 cm^{-3}, and the mean electron energy equal to 1 eV.

Presuming disappearance of the metastable atom density, i.e., $n_m = 0$ has been arranged at the cathode, whilst a predominant field-driven flux shut to the cathode the expression $\partial D_+ n_+ / \partial x = 0$ has been used for the positive ion density $\forall t > 0$. At the anode, the metastable atoms and the electron density are supposed to be zero. The electron flux separating the cathode is calculated by the expression $\varphi_e(x = 0, t) = -\gamma\varphi_+(x = 0, t)\ \forall t > 0$, the mean electron energy is assumed to be 5 eV at the cathode [36] and the gas temperature is equal to 273 K in the discharge.

2.2 Numerical method

For the metastable atom and Poisson equations a finite difference method has been employed. The transport equations of the electron energy, electron and ion are also discretized spatially with the finite difference technique. In this method the exponential scheme has been employed into account [37, 39–42]. The discretization of the time terms by the right position of the finite difference technique has been used. Consequently, every discretized equation is defined as a tridiagonal matrix, which is solved by Thomas's algorithm.

3. Results and discussion of argon discharge

In this part, we will analysis the spatiotemporal evolution of the abnormal glow discharge in the existence of metastable atom density. The gas pressure is 133.32 Pa. The neutral species density is computed from the temperature and gas pressure with the ideal gas law. The constant value for the secondary electron yield is 0.06 [36].

The applied potential at the cathode is −400 V. The uniform subdivision of the space interval in 250 elements and a time step Δt = 10 ps, have been utilized.

Figure 1 shows the temporal progression of the potential (**Figure 1(a)**), concentration of electrons (**Figure 1(b)**), concentration of ions (**Figure 1(c)**), concentration of metastable atoms (**Figure 1(d)**), the electric field (**Figure 1(e)**) and the mean electron energy (**Figure 1(f)**). We remark that the discharge is distinguished by three zones: the first one busied the time simulation amidst in 10^{-11} and 9×10^{-6} s, the second one took the time simulation inter-time of 9×10^{-6} and 3×10^{-5} s, the final zone busied the time simulation amidst in 3×10^{-5} and 3.6×10^{-5} s.

For the first zone, we comment that the metastable atom, electrons and ion concentrations are almost identical. Therefore, the net space charge concentration is unimportant. The electric potential is distinguished by the Laplace form due to the net space charge concentration that is exist. Consequently, the electric field is seemingly constant, besides that the mean electron energy is constant.

In the second zone, we remark a pseudo emergence of the cathodic region, this is characterized by a significance of ion concentration and unimportant of the electron

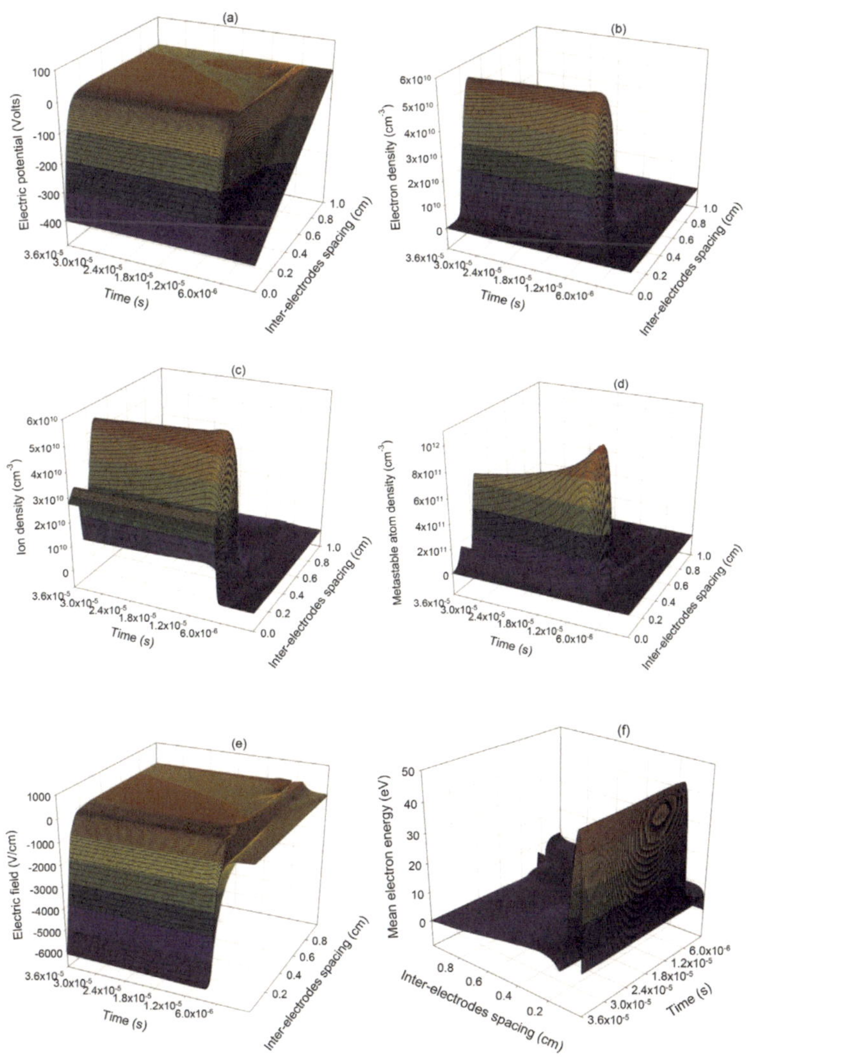

Figure 1.
Argon spatiotemporal distributions of electric potential (a), electrons volume number (b), ions volume number (c), metastable atom density (d), electric field (e), and mean electron energy (f) at 400 V.

concentration. This is explicated by the velocity of the electron species, which speed a more than the ion species and displaced quickly starting at the cathodic region. Consequently, the amount (n_e-n_+) is considerable which influence directly the electric potential as consequence the important chute that is present. Inevitably, the electric field is intense. The last earns the electron species an important energy. We comment that the metastable atom concentration is important. This discharge is sustained by the secondary emission coefficient as well as the existence of metastable atom concentration. Sooner than $t = 3 \times 10^{-5}$ s we remark a pseudo emergence of the negative glow region, where it is typified by the similar electron and ion concentrations. Consequently the net space charge concentration is slight. Hence, the electric potential and the electric field are constants. Automatically, the metastable atoms concentration is diminished.

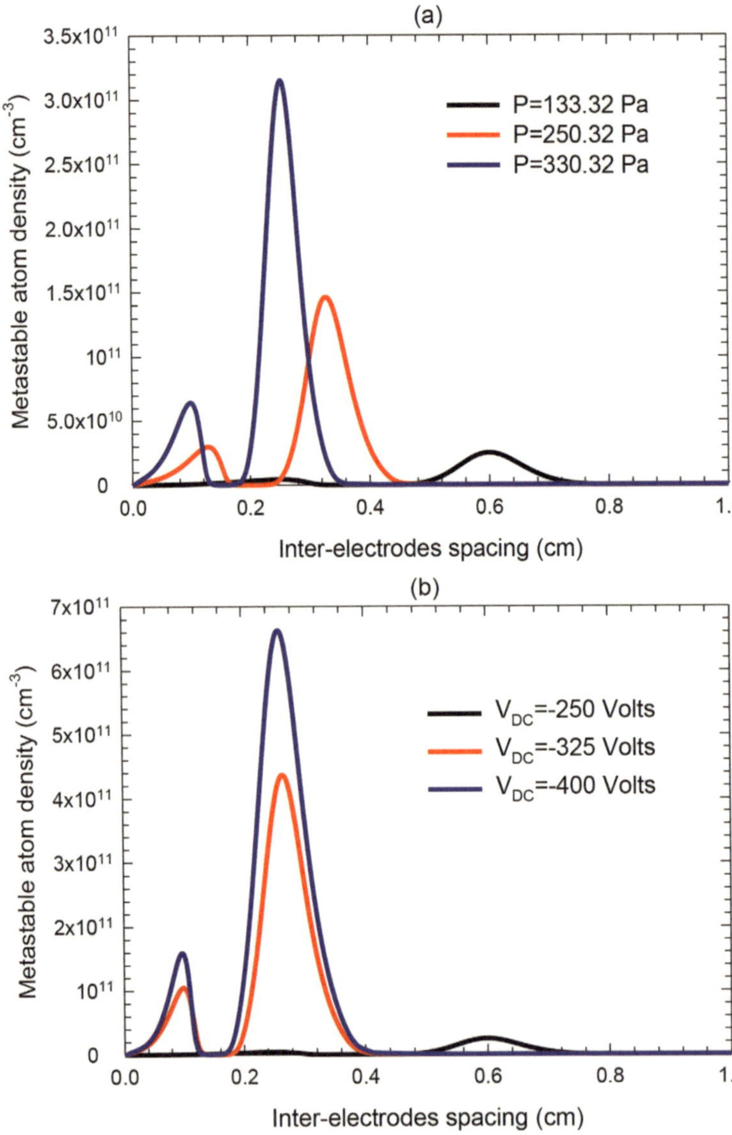

Figure 2.
Metastable atom densities as a function of pressure (a) and as a function of voltage (b) in the stationary state in Argon gas.

In the last zone, we remark three dissimilar regions: the anode and the plasma regions and the cathode region. The anode region is characterized by the ion concentration that is less important compared to the electron concentration. In this zone we remark that the convergence of each physical characteristics of the discharge are reached at the time 3.6×10^{-5} s.

3.1 Influence of the voltage and gas pressure

In this part, we will study the influence of the voltage and gas pressure on the argon discharge. So, the potential at the cathode is taking of 250 V and we will alter the gas pressure. For the influence of the applied potential on the discharge, we take the gas pressure at 133.32 Pa. bulk.

Figure 2(a) shows the metastable atoms concentration plots depending on the pressure in the study state. The metastable atoms concentration augments with increasing pressure. For elevated pressure the gas density augments, which the electron diffusion coefficient turns out to be fewer and the bulk of the plasma rises which the both sheaths of the anode and cathode turn out to be small. These circumstances of the charged particle manipulate on the metastable atoms behavior in the study stated, i.e., the cathodic region is overflowing with the electron and metastable atom concentrations, which go faster the ion species in the existence of the electric field.

Figure 2(b) shows the metastable atoms concentration graphs depending on the potential in the study state. For elevated potential the excitation and ionization processes increase, and the charged particle turns out to be raised in the stationary state. Consequently, the metastable atom concentration graphs become growing.

The greatest of the metastable atom concentration varies amidst in 2.47×10^{10} and 6.63×10^{11} cm^{-3}. We judged these results with calculated values established in the literature [9, 10, 12, 15]. Depending on the discharge circumstances, all these value varies amidst in 2×10^{10} and 5×10^{13} cm^{-3}. Consequently, our calculated values something like in the exact range of the size order. We find that the values of

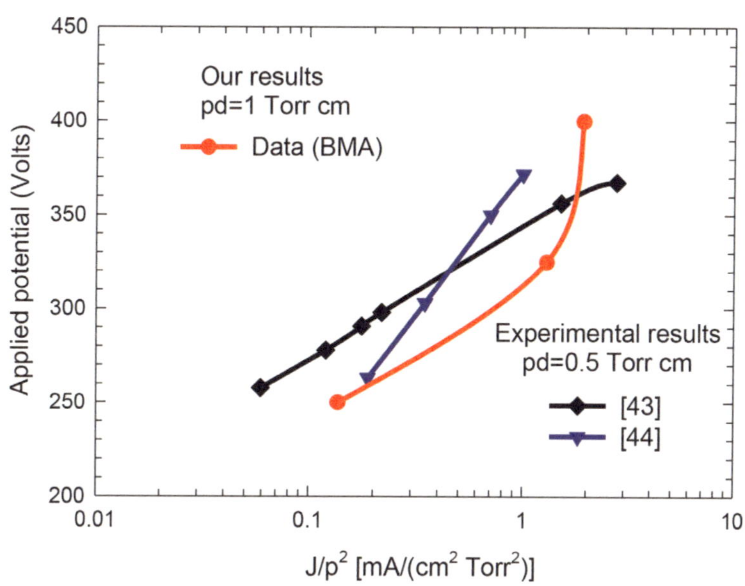

Figure 3.
Comparison between the results obtained by our calculation using a database of BMA for pd = 1 Torr cm and those given by experimental for pd = 0.5 Torr cm in Argon gas.

the current densities are 0.137, 0.508 and 0.843 mA/cm^2 related to the pressures of 133.32, 250.32 and 330.32 Pa, in that order.

Figure 3 represents the current-voltage properties in the stationary state. The results attained beginning of the database of BMA evaluated to those obtained by experimental method [43, 44]. We find that the results get beginning of the database of the BMA are in excellent conformity judgment against to those experimental results [43, 44].

4. Validity of the model

Figure 4 demonstrates the comparison amidst in our results and those given by Becker et al. [30] (**Figure 4(a)**) ion and electron concentrations, (**Figure 4(b)**) electric field and electric potential, (**Figure 4(c)**) metastable atom concentration and (**Figure 4(d)**) mean electron energy. This figure substantiates the validity of our 1D code. The main dissimilarity amidst in those given by Becker et al. and our results are pointing up in the **Table 2**.

We find that the similar discharge has been studied by Fiala et al. [45], where the hybrid model has been employed in two dimensional configurations. We find that the results got by Fiala et al., it was approximately indistinguishable to our results. In exacting, the maximum of particle concentrations is 1.1×10^9 cm^{-3} and the electric field at the cathode is 675 V/cm for applying voltage that is equal to 126.3 V. Consequently, the hybrid model is equivalent to our fluid model in the presence of the metastable atom concentrations of these discharge circumstances. Besides our model identifier both properties of the discharge, the mean electron energy and the metastable atom concentrations.

5. Influence for input data of argon abnormal glow discharge

In this part, we will show the properties of the argon plasma discharge in the case of entering data computed by BOLSIG+ software [46]. We remind that the preceding results are obtained with enter data computed by multiterm estimation of the Boltzmann equation. We remind again that the preceding results are identical when are calculated exclusive of the rate coefficients K_m^o and K_m^{io}. The exclusion is the metastable atom concentration, which is prejudiced by these coefficients, i.e., the stepwise ionization processes is insignificant, evaluated to the both ionization of the chemo-ionization and ground state atoms processes. Consequentially, we can calculate the properties of the argon abnormal glow discharge exclusive of K_m^o and K_m^{io} coefficients. The secondary electron emission coefficient is 0.06. The applied voltage at the cathode is −250 V. The gas pressure is 133.32 Pa.

Figure 5 shows the particle density distributions (**Figure 5(a)**), metastable atom concentration (**Figure 5(b)**) and mean electron energy (**Figure 5(c)**) in the stationary state. When we compare between the results given by **Figures 4** and 5, we conclude clearly the influence of entering data of BOLSIG+ on the characteristics of the argon abnormal glow discharge. In exacting the cathodic region illustrated in the **Figure 4(a)** is totally dissimilar to the cathodic region illustrated in the **Figure 5(a)**. We remind that the similar discharge circumstances (voltage, pressure, etc.) are used in two and multi terms approximations. But, the results are actually dissimilar.

Table 3 gives the main dissimilarity between entering both data of multiterm approximation and of BOLSIG+.

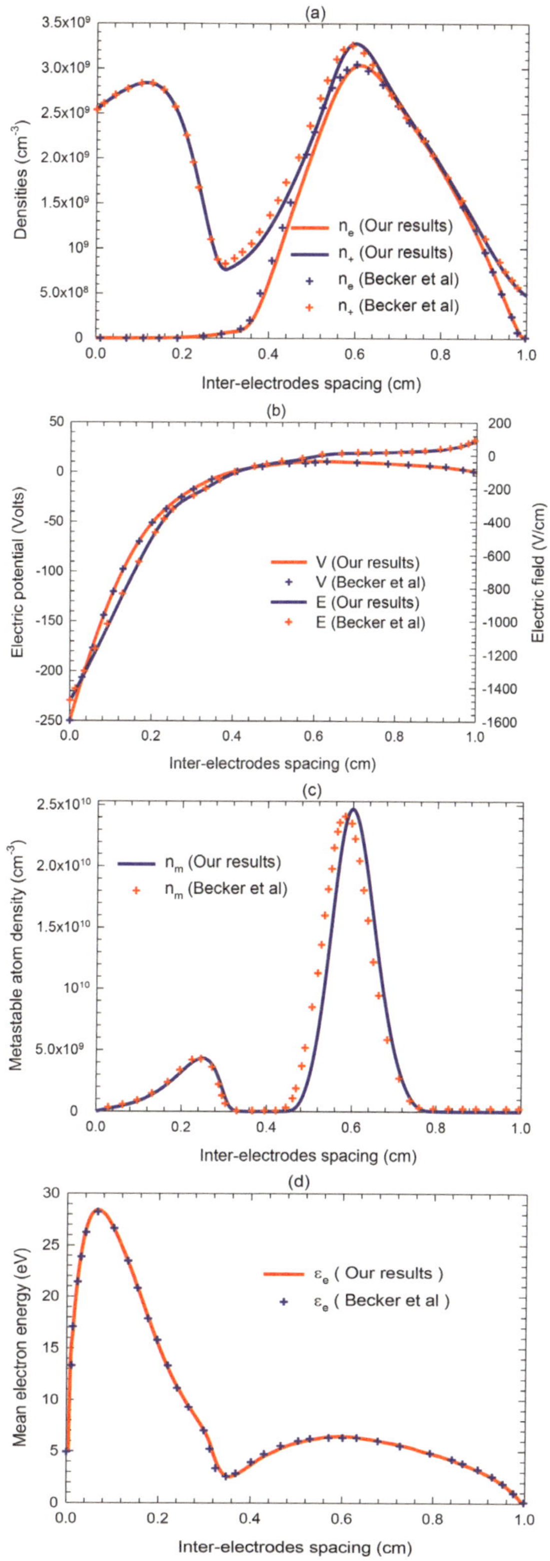

Figure 4.
Comparison between our results and those given by Becker et al.: (a) densities, (b) electric potential and electric field, (c) metastable atom density, and (d) mean electron energy in the stationary state in Argon gas.

Becker et al [30]	Our results
Stabilized finite element method for the transport and energy equations are used.	Scharfetter and Gummel scheme is used
The applied potential at the cathode is $V_{DC}(1-e^{-t/\tau})$ and $\tau = 1$ ns	The applied potential at the cathode is V_{DC}
The steady state is attained at 10^{-4} s	The steady state is attained at 3.6×10^{-5}s
The computations were performed on a 3GHz CPU and typically took about four hours to reach the steady state.	The calculations are carried out in a personal computer. It typically takes about five hours to reach the steady state
The parametric studies are not presented	The parametric studies of gas pressure and applied potential are presented

Table 2.
The major differences between our results and those given by Becker et al.

Figure 6 shows the comparison amidst of the experimental results [47, 48] and our results got beginning database of BOLSIG+ code for pd = 2 Torr cm. We remind that the experimental results [47, 48] are given for the diameter of the electrode equal to 8 cm and the inter-electrodes spacing equal to 1 cm. We remind again that the results obtained by Ref. [47] its approximately dissimilar to the results obtained by Ref. [48] due to the experimental circumstances of the secondary electron emission coefficient. We remark that our results got from database of BOLSIG+ code are in excellent accord with the experimental results [47].

5.1 Effect of the metastable lifetime on the characteristics of argon abnormal glow discharge

We remind that the preceding computations are effected through the metastable lifetime equal to 1 μs [30], we remind that this value has been proposed by Becker et al. [30]. **Figure 7** represents the effect of the metastable lifetime on the curve of the metastable atom concentration in the study state. For this reason we have used a value of metastable lifetime equal to 56 s of the theories [49] and an experimental value equal to 38 s [50]. We remark that the most of the metastable atom concentration augments from 7.76×10^9 to 2.249×10^{11} cm^{-3}. We note that this effect is noticed just for metastable atom concentration and all characteristic of argon abnormal glow discharge rest unmoved in the study stated. As a consequence, the utilization of the experimental or artificial value of the metastable lifetime has an inconsequential of the abnormal glow discharge characteristics.

6. Characteristics of the neon discharge through entering data of the BOLSIG+

In this part, we will investigate the characteristics of the neon abnormal glow discharge through entering data of the BOLSIG+ code. The gas pressure is 3 Torr. The applied voltage is 300 V. The secondary electron emission coefficient is 0.26 [51]. Additional parameters are declared in the Appendix A. **Figure 8** shows the curves of ion and electron densities (**Figure 8(a)**), metastable atom concentration

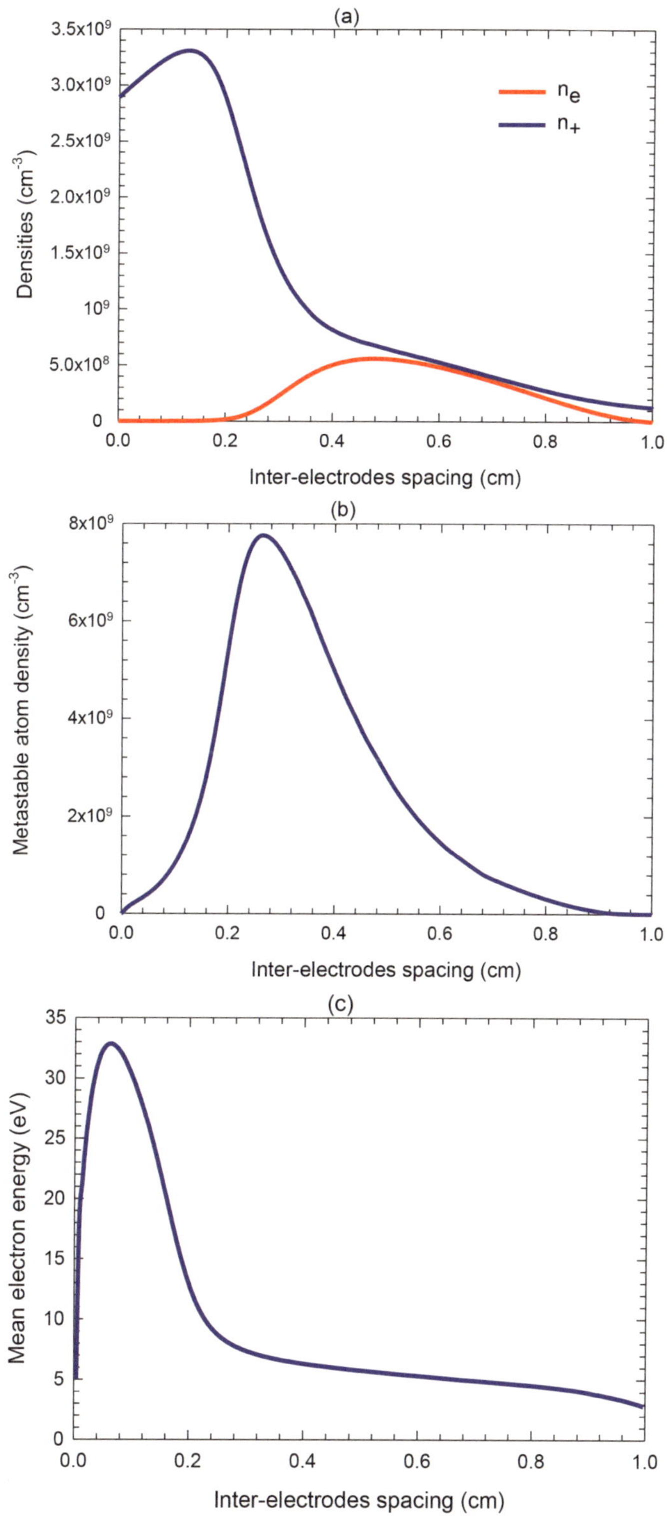

Figure 5.
Argon spatial distributions of (a) particledensities, (b) metastable atom density, and (c) mean electron energy in the stationary state for input data of BOLSIG+.

Input-Data (multi term approximation)	Input-Data (BOLSIG+)
Boltzmann equation is solved in the 8-term approximation	Boltzmann equation is solved in the two-term approximation
The activated energy of K_o^m is ≈ 3 eV	The activated energy of K_o^m is << 3 eV
The activation energy of K_o^{io} is ≈ 5 eV	The activated energy of K_o^{io} is << 5 eV
Electron mobility is almost great	The electron mobility is great

Table 3.
The major differences between input Data of BOLSIG+ and those given by multiterm approximation.

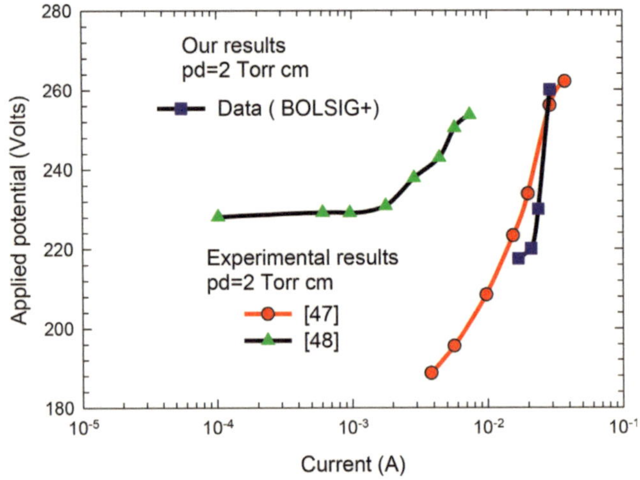

Figure 6.
Comparison between the results obtained by our calculation using database of BOLSIG+ software for pd = 2 Torr cm and those given by experimental for pd = 2 Torr cm in Argon gas.

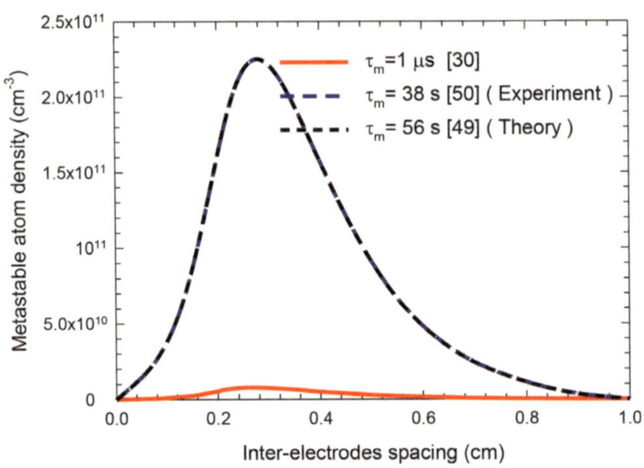

Figure 7.
Influence of the metastable lifetime on the Argon spatial distribution of metastable atom density in the steady state.

(**Figure 8(b)**) and current density (**Figure 8(c)**) depending of the applied voltage in the stationary state. We note that the greatest of the metastable atom concentration is equal to 1.615×10^{11} cm^{-3}. The value of neon current density is 0.1851 mA/cm^2. The electric field at the cathode achieves the value of 1696.08 V/cm. We note

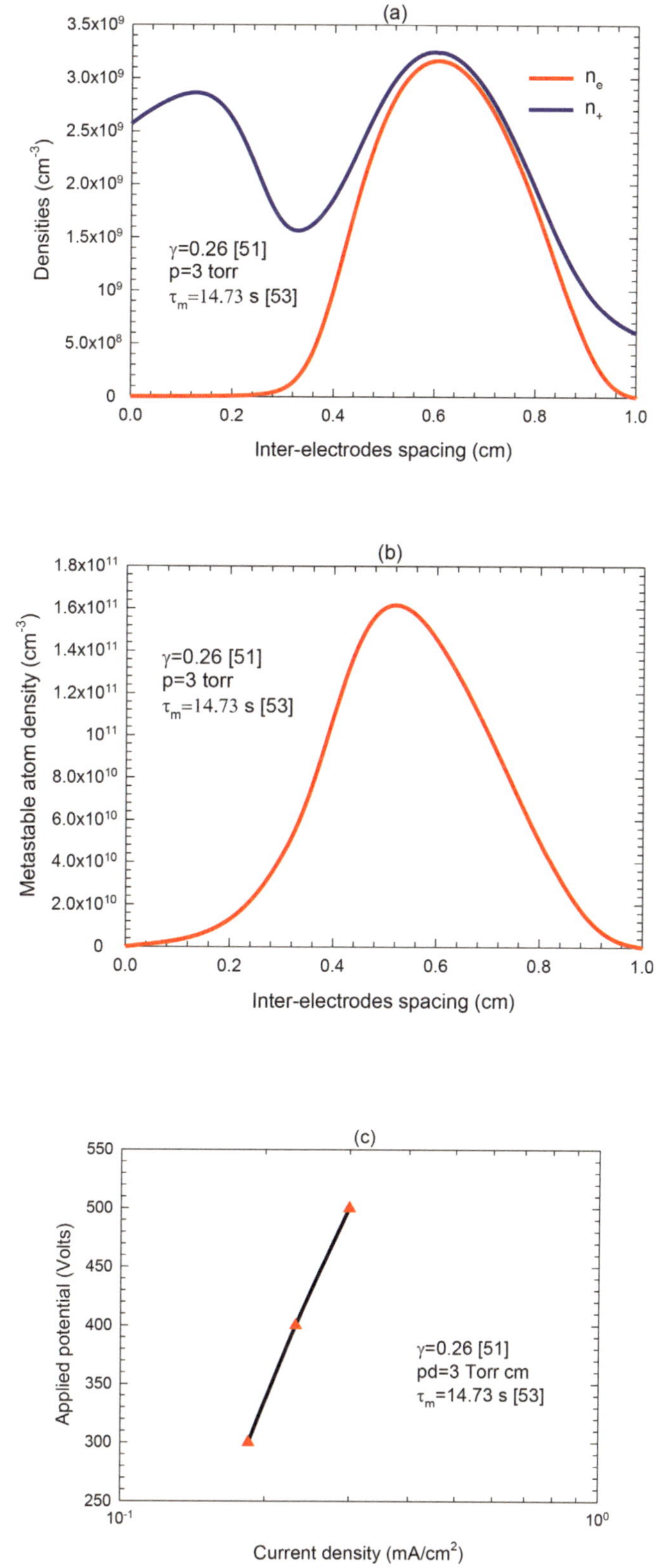

Figure 8.
Neon spatial distributions of (a) particle densities, (b) metastable atom density, and (c) current density as a function of applied potential in the stationary state.

that the most of the mean electron energy is 46.30 eV in the cathodic region. As a consequence, the fluid model by injecting the metastable atom concentrations is extremely important for investigating the abnormal glow discharge properties of several pure gases and mixed gas.

7. Conclusion

In order to study the effect of metastable atom density, a second order fluid electric model has been used in the case of a DC low-pressure Ar and Ne abnormal glow discharges. The Poisson equation for the potential and electric field is joined to the first three moments of the Boltzmann's conservation equations ignoring inertia of the charged particles. In the framework of the local energy approximation, the basic data employed in this chapter are calculated by Becker et al. in the case of multiterm estimation of Boltzmann equation (BMA) [30] and from BOLSIG+ software. The task of metastable atom concentration in the discharge is obvious for study into side of plasma glow discharge for several pure gases and mixed gas. We note that the abnormal glow discharge is sustained by secondary electron emission coefficient and the existence of the metastable atom concentration in this particular discharge.

A. Appendix A

The drift velocity of positive ion neon gas is: $w_+ = (11.27E/n)/(1 + 0.01288E/n)^{0.5}$ (m/s) [52] where E/N is in Td. The metastable lifetime of neon gas is equal to 14.73 s [53]. The diffusion coefficient of metastable atoms is $D_m = 150$ cm^2 s^{-1} Torr [54]. The energy loss of an excited atom is 16.6 eV. The Ionization energy of neon gas is equal to 21.56454 eV. The energy loss (P^{ec}) per electron due to elastic collision of electrons with the background gas is calculated according to [31, 55]. The rate coefficient of chemo-ionization processes is $K_{ci} = 3.6 \times 10^{-10}$ cm^3 s^{-1} for T = 310 K [56]

Acknowledgements

The authors express their gratitude to Markus M Becker (Greifswald-Germany) for helpful discussions.

Author details

Abdelaziz Bouchikhi
Faculty of Technology, Department of Electrical Engineering, University of Saïda, Saïda, Algeria

*Address all correspondence to: bouchikhiabdelaziz1@yahoo.fr

References

[1] Ono S, Kato H, Tell S. Simulation and experiment of low-pressure oxygen-nitrogen mixture gas glow discharge plasma. Combustion Science and Technology. 1998;**133**:151

[2] Khomich VA, Ryabtsev AV, Didyk EG, Zhovtyansky VA, Nazarenko VG. Numerical simulation of atomic nitrogen formation in plasma of glow discharge in nitrogen-argon mixture. Technical Physics Letters. 2010;**36**:918

[3] Ponduri S, Becker MM, Welzel S, van de Sanden MCM, Loffhagen D, Engeln R. Fluid modelling of CO_2 dissociation in a dielectric barrier discharge. Journal of Applied Physics. 2016;**119**:093301

[4] Baadj S, Harrache Z, Belasri A. Electrical and chemical properties of XeCl*(308 nm) exciplex lamp created by a dielectric barrier discharge. Plasma Physics Reports. 2013;**39**:1043

[5] Li Q, Zhu X, Li J, Pu Y. Role of metastable atoms in the propagation of atmospheric pressure dielectric barrier discharge jets. Journal of Applied Physics. 2010;**107**:043304

[6] Eckstein EW, Coburn JW, Kay E. Diagnostics of an r.f. sputtering glow discharge - correlation between atomic absorption and mass spectrometry. International Journal of Mass Spectrometry and Ion Physics. 1975;**17**:129

[7] Uzelac NI, LeisF F. Measurement of gas temperatures and metastable state densities in a microwave boosted glow discharge using a diode laser. Spectrochimica Acta. 1992;**47B**:877

[8] Browne PG, Dunn MH. Metastable densities and excitation processes in the He-Cd laser discharge. Journal of Physics B. 1973;**6**:1103

[9] Smith RL, Serxner D, Hess KR. Assessment of the relative role of Penning ionization in low-pressure glow discharges. Analytical Chemistry. 1989; **61**:1103

[10] Strauss JA, Ferreira NP, Human HGC. An investigation into the role of metastable argon atoms in the afterglow plasma of a low pressure discharge. Spectrochimica Acta. 1982;**37B**:947

[11] Ferreira NP, Strauss JA, Human HGC. Developments in glow discharge emission spectrometry. Spectrochimica Acta. 1982;**37B**:273

[12] Ferreira CM, Loureiro J, Ricard A. Populations in the metastable and the resonance levels of argon and stepwise ionization effects in a low-pressure argon positive column. Journal of Applied Physics. 1985;**57**:82

[13] Ferreira CM, Ricard A. Modelling of the low-pressure argon positive column. Journal of Applied Physics. 1983;**54**:2261

[14] Hardy KA, Sheldon JW. Metastable atom density in helium, neon, and argon glow discharges. Journal of Applied Physics. 1982;**53**:8532

[15] Kubota T, Morisaki Y, Ohsawa A, et al. The axial distributions of optical emission and metastable density: comparison between experiments with DC and RF helium glow discharges. Journal of Physics D. 1992;**25**:613

[16] Lymberopoulos DP, Economou DJ. Fluid simulations of glow discharges: Effect of metastable atoms in argon. Journal of Applied Physics. 1993;**73**:3668

[17] Phelps AV, Molnar JP. Lifetimes of metastable states of noble gases. Physics Review. 1953;**89**:1202

[18] Phelps AV. Absorption studies of helium metastable atoms and molecules. Physics Review. 1955;**99**:1307

[19] Ellis E, Twiddy ND. Time-resolved optical absorption measurements of excited-atom concentrations in the argon afterglow. Journal of Physics B. 1969;**2**:1366

[20] Copley GH, Lee CS. Electron excitation and deexcitation coefficients for the $^3P_2$, $^3P_1$, $^3P_0$, and $^1P_1$ levels of argon. Canadian Journal of Physics. 1975;**53**:1705

[21] Tachibana K. Excitation of the $1s_5$,$1s_4$, $1s_3$, and $1s_2$ levels of argon by low-energy electrons. Physical Review A. 1986;**34**:1007

[22] Kolts JH, Setser DW. Decay rates of $Ar(4s,^3P_2)$, $Ar(4s',^3P_0)$, $Kr(5s,^3P_2)$, and $Xe(6s,^3P_2)$ atoms in argon. The Journal of Chemical Physics. 1978;**68**:4848

[23] Den Hartog EA, O'Brian TR, Lawler JE. Electron temperature and density diagnostics in a helium glow discharge. Physical Review Letters. 1989;**62**:1500

[24] Den Hartog EA, Doughty DA, Lawler JE. Laser optogalvanic and fluorescence studies of the cathode region of a glow discharge. Physical Review A. 1988;**38**:2471

[25] Fedoseev AV, Sukhinin GI. Influence of metastable argon atoms and dust particles on gas discharge plasma. Ukrainian Journal of Physics. 2011;**56**:1272

[26] Shumova VV, Polyakov DN, Vasilyak LM. Effect of metastable neon atoms in a positive column of glow discharge with dust particles. Plasma Sources Science and Technology. 2014; **23**:065008

[27] Vasilyak LM, Polyakov DN, Shumova VV. Glow discharge positive column with dust particles in neon. Contributions to Plasma Physics. 2013; **53**:432

[28] Sukhinin GI, Fedoseev AV, Antipov SN, et al. Dust particle radial confinement in a dc glow discharge. Physical Review E. 2013;**87**:013101

[29] Alili T, Bouchikhi A, Rizouga M. Investigations of argon and neon abnormal glow discharges in the presence of metastable atom density with fluid model. Canadian Journal of Physics. 2016;**94**:731

[30] Becker MM, Loffhagen D, Schmidt W. A stabilized finite element method for modeling of gas discharges. Computer Physics Communications. 2009;**180**:1230

[31] Kolokolov NB, Kudrjavtsev AA, Blagoev AB. Interaction processes with creation of fast electrons in the low temperature plasma. Physica Scripta. 1994;**50**:371

[32] Sigeneger F, Winkler R. Nonlocal transport and dissipation properties of electrons in inhomogeneous plasmas. IEEE Transactions on Plasma Science. 1999;**27**:1254

[33] Belenguer P, Boeuf JP. Transition between different regimes of rf glow discharges. Physical Review A. 1990;**41**: 4447

[34] Donkó Z. Hybrid model of a rectangular hollow cathode discharge. Physical Review E. 1998;**57**:7126

[35] Marié D, Kutasi K, Malovié G, et al. Axial emission profiles and apparent secondary electron yield in abnormal glow discharges in argon. European Physical Journal D: Atomic, Molecular, Optical and Plasma Physics. 2002;**21**:73

[36] Phelps A, Petrović Z. Cold-cathode discharges and breakdown in argon: Surface and gas phase production of secondary electrons. Plasma Sources Science and Technology. 1999;**8**:R21

[37] Bouchikhi A. Two-dimensional numerical simulation of the DC glow discharge in the normal mode and with

Einstein's relation of electron diffusivity. Plasma Science and Technology. 2012;**14**:965

[38] Becker MM, Loffhagen D. Enhanced reliability of drift-diffusion approximation for electrons in fluid models for nonthermal plasmas. AIP Advances. 2013;**3**:012108

[39] Scharfetter DL, Gummel HK. Large-signal analysis of a silicon read diode oscillator. IEEE Transactions on Electron Devices. 1969;**16**:64

[40] Bouchikhi A, Hamid A, Flitti A, et al. The application of the 2 order fluid model for the townsend's discharge study. Acta Electrotehnica. 2008;**48**:404

[41] Bouchikhi A, Hamid A. 2D DC subnormal glow discharge in argon. Plasma Science and Technology. 2010; **12**:59

[42] Stankov MN, Petković MD, Marković VL, et al. Numerical modelling of DC argon glow discharge at low pressure without and with Ar ($^3P_2$) metastable state. Romanian Journal of Physics. 2014;**59**:328

[43] Stefanović I, Petrovic´ ZL. Volt ampere characteristics of low current DC discharges in Ar, H_2, CH_4 and SF_6. Japanese Journal of Applied Physics Part 1. 1997;**36**:4728

[44] Rozsa K, Gallagher A, Donkó Z. Excitation of Ar lines in the cathode region of a DC discharge. Physical Review E. 1995;**52**:913

[45] Fiala A, Pitchford LC, Boef JP. Two-dimensional, hybrid model of low-pressure glow discharges. Physical Review E. 1994;**49**:5607

[46] Hagelaar G, Pitchford L. Solving the Boltzmann equation to obtain electron transport coefficients and rate coefficients for fluid models. Plasma Sources Science and Technology. 2005; **14**:722

[47] Petrović Z, Jelenković B, Phelps A. Communication Privée; 1994

[48] Jelenković B, Rózsa K, Phelps A. Oscillations of low-current electrical discharges between parallel-plane electrodes. II. Pulsed discharges in H_2. Physical Review E. 1993;**47**:2816

[49] Small-Warren NE, Chow-Chiu LY. Lifetime of the metastable $3P_2$ and $3P_0$ states of rare-gas atoms. Physical Review A. 1975;**11**:1777

[50] Katori H, Shimizu F. Lifetime measurement of the $1s_5$ metastable state of argon and krypton with a magneto-optical trap. Physical Review Letters. 1993;**70**:3545

[51] Chapman B. Glow Discharge Processes. New York: John Wiley Sons; 1980

[52] Frost LS. Effect of variable ionic on mobility on ambipolar diffusion. Physics Review. 1957;**105**:354

[53] Zinner M, Spoden P, Kraemer T, et al. Precision measurement of the metastable $^3P_2$ lifetime of neon. Physical Review A. 2003;**67**:010501(R)

[54] Ricard A. Evolution de la densité des atomes métastables du néon formés dans une décharge à courant continu de faible intensité. JPHS. 1969;**30**:556

[55] Gaens W, Bogaerts A. Kinetic modelling for an atmospheric pressure argon plasma jet in humid air. Journal of Physics D: Applied Physics. 2014;**47**: 079502

[56] Sheverev VA, Stepaniuk VP, Lister GG. Chemi-ionization in neon plasma. Journal of Applied Physics. 2002;**92**: 3454

Section 3

Spintronics

Chapter 7

Application of Spin-Orbit Coupling in Exotic Graphene Structures and Biology

Richard Pinčák and Erik Bartoš

Abstract

An important measurable quantity in the carbon nanostructures, including the nanotubular part of the graphitic wormhole, is the spin-orbit coupling. We will present in this chapter spin-orbit coupling for the fermions located in exotic graphene structures as is graphene wormhole and also in biological systems. Considering this influence, the two-component Dirac equation is changed into the usual four-component form. As a consequence, the chiral fermions should be detected close to the wormhole bridge. We will show that the smaller is the radius of the wormhole bridge, the stronger this effect should be. Finally, we will describe the role of spinor fields in the time series of genetic code. The reversed transcription process of the gene expression could be defined by a moduli state space model of a coupling spinor field between the gene of a viral particle and the host cell. As a general result, all states of codon can be computed by the Chern-Simons 3-forms.

Keywords: spinor network structure, spin orbit coupling, Chern-Simons fields, graphene wormhole, genetic code

1. Graphitic wormhole

The investigation of unique chemical and mechanical properties of nanostructures, e.g., fullerene, graphene, and nanotubes, promises a wide application in many technical areas. The electronic properties of the nanostructures are basically defined by their hexagonal carbon lattice structure and its variations. New promising results are expected with the preparation of more complicated forms as a wormhole. The wormhole is usually composed of two different kinds of nanostructure: two graphene sheets are connected together with the help of a connecting nanotube [1] (**Figure 1**). This is achieved by a supply of two sets of six heptagonal defects onto both sides of the given nanotube. There exists the restrictions on the form of the nanotube—the chirality must be $(6n, 6n)$ armchair or $(6n, 0)$ zigzag and a radius of the nanotube is larger than its length.

The metric tensor of the wormhole is given by

$$g_{\mu\nu} = \Lambda^2(r_\pm)\begin{pmatrix} 1 & 0 \\ 0 & r_\pm^2 \end{pmatrix}, \quad \Lambda(r_\pm) = (a/r_\pm)^2\theta(a - r_\pm) + \theta(r_\pm - a), \tag{1}$$

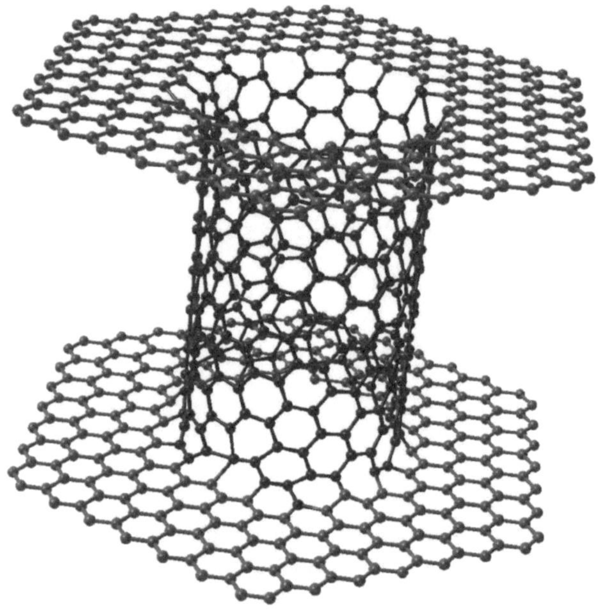

Figure 1.
Schematic representation of graphitic wormhole consisting from two graphene sheets connected together with the help of a nanotube.

where θ is the Heaviside step function, r_- and r_+ are the polar coordinates of lower and upper graphene sheets, and $a = \sqrt{r_- r_+}$ is the radius of the wormhole.

1.1 Electronic structure

We consider the continuum gauge field theory, i.e., at each point of a molecular surface, we take into account an influence of different gauge fields that enter the Dirac-like equation for an electron

$$i v_F \sigma^\mu \left[\partial_\mu + \Omega_\mu - i a_\mu - i a_\mu^W - i A_\mu\right]\psi = E\psi, \tag{2}$$

with σ^α matrices as the Pauli matrices, the Fermi velocity v_F, the spin connection

$$\Omega_\mu = \frac{1}{8}\omega_\mu^{\alpha\beta}\left[\sigma_\alpha, \sigma_\beta\right], \tag{3}$$

and the covariant derivative $\nabla_\mu = \partial_\mu + \Omega_\mu$. The gauge fields a_μ, a_μ^W are caused by the presence of the defects, and by rotational symmetry, the gauge field A_μ characterizes the possible magnetic field.

In the case of the wormhole with the metric Eq. (1), the effective flux caused by the presence of the defects is included in the gauge field a_μ, and for the particular polar components, it has the values $a_\varphi = \frac{3}{2}$, $a_r = 0$ for two possibilities: the first corresponds to the chiral vector with the form $(6n, 6n)$, the second corresponds to the chiral vector with the form $(6n, 0)$ and n divisible by 3. In the case of chiral vector of the form $(6n, 0)$, where n is not divisible by 3, the components of the gauge field are $a_\varphi = \frac{1}{2}$, $a_r = 0$. Regarding that the components of the spin connection are

$$\Omega_\varphi = -\frac{\mathrm{i}}{2}\sigma_3\left(r\frac{\Lambda'(r)}{\Lambda(r)}+1\right), \quad \Omega_r = 0, \tag{4}$$

and after the substitution into Eq. (2), we get the equation

$$\mathrm{i}v_F\sigma^\mu\left(\partial_\mu + \Omega_\mu \mp \mathrm{i}\, a_\mu\right)\psi^\pm = \varepsilon\psi^\pm, \tag{5}$$

where each sign corresponds to a different Dirac point

$$\begin{aligned} -\mathrm{i}v_F\left(\partial_r + \frac{1}{r}\mathrm{i}\partial_\theta \mp \frac{a_\varphi}{r} + \frac{1}{2r}\right)\psi_B^\pm &= \varepsilon\psi_A^\pm, \\ -\mathrm{i}v_F\left(\partial_r - \frac{1}{r}\mathrm{i}\partial_\theta \pm \frac{a_\varphi}{r} + \frac{1}{2r}\right)\psi_A^\pm &= \varepsilon\psi_B^\pm \end{aligned} \tag{6}$$

for $r \geq a$ and

$$\begin{aligned} \mathrm{i}v_F\left(\tfrac{r}{a}\right)^2\left(\partial_r - \frac{1}{r}\mathrm{i}\partial_\theta \pm \frac{a_\varphi}{r} - \frac{1}{2r}\right)\psi_B^\pm &= \varepsilon\psi_A^\pm, \\ \mathrm{i}v_F\left(\tfrac{r}{a}\right)^2\left(\partial_r + \frac{1}{r}\mathrm{i}\partial_\theta \mp \frac{a_\varphi}{r} - \frac{1}{2r}\right)\psi_A^\pm &= \varepsilon\psi_B^\pm \end{aligned} \tag{7}$$

for $0 < r \leq a$. For $r \geq a$, the solution is

$$\begin{aligned} \psi^\pm = \begin{pmatrix} \psi_A^\pm(r,\varphi) \\ \psi_B^\pm(r,\varphi) \end{pmatrix} &= c_1\begin{pmatrix} J_{n\mp a_\varphi-1/2}(kr) \\ -\mathrm{i}\ \mathrm{sgn}\ \varepsilon J_{n\mp a_\varphi+1/2}(kr) \end{pmatrix} \\ &+ c_2\begin{pmatrix} Y_{n\mp a_\varphi-1/2}(kr) \\ -\mathrm{i}\ \mathrm{sgn}\ \varepsilon Y_{n\mp a_\varphi+1/2}(kr) \end{pmatrix}, \end{aligned} \tag{8}$$

where the energy $\varepsilon = \pm v_F k$, $J_n(x)$ and $Y_n(x)$ are the Bessel functions.

The zero modes solve the Dirac equation for zero energy. If one choose the component $\psi_A^\pm$ of the solution to be equal to zero, one get from (6) and (7)

$$\left(\partial_r - \frac{1}{r}\mathrm{i}\partial_\theta \mp \frac{a_\varphi}{r} + \frac{1}{2r}\right)\psi_B^\pm = 0 \tag{9}$$

for $r \geq a$ and

$$\left(\partial_r - \frac{1}{r}\mathrm{i}\partial_\theta \pm \frac{a_\varphi}{r} - \frac{1}{2r}\right)\psi_B^\pm = 0 \tag{10}$$

for $0 < r \leq a$. For ψ_B^- and the value $a_\varphi = \frac{3}{2}$, the solution is

$$\psi_B^-(r,\varphi) \sim r^{-n-2}e^{\mathrm{i}n\varphi} \tag{11}$$

for $r \geq a$ and

$$\psi_B^-(r,\varphi) \sim r^{-n+2}e^{\mathrm{i}n\varphi} \tag{12}$$

for $0 < r \leq a$. For both cases, it is normalizable only for $n = 0$, and so this is the only solution. In a similar way, we can calculate the zero modes for the component ψ_B^+. For the value $a_\varphi = \frac{1}{2}$, possible solutions are not strictly normalizable, and the zero modes exist only for the case of the connecting nanotube being armchair or zigzag with the chiral vector $(6n, 0)$, n divisible by 3. In other cases the zero modes do not exist.

Recently, in work [2] some peculiarities in the bilayer graphene were analytically predicted. A possible indication of the wormhole could be found in [3, 4], where a new type of zero modes is investigated. These zero modes could be the zero modes studied in this subsection applied to the case of the smallest wormhole.

1.2 Case of massive fermions

Up to now we supposed that the fermions appearing in the Dirac equation have the zero mass or that the mass is very small in comparison with their energy, but in [5, 6] it was shown that the Fermi velocity needs to be renormalized due to the elasticity and deformations in a graphene. In our case of the graphitic wormhole, including big deformations, the velocity of fermions close to the wormhole bridge could achieve such values that the relativistic effects can appear or break off the symmetry [7] and the mass of fermions would be non-negligible. The radius of the wormhole and its bridge is very small in comparison with the size of the upper and the lower graphene sheet (**Figure 2**) and by folding the sheet into a tube they acquire nonzero effective mass as they move along the tube axis. This change of the space topology of graphene from 2D to 1D space compactification is similar to the string theory compactification, and we can imagine a wormhole connecting nanotubes as 1D object.

To include the mass into the Dirac Eq. (2), one can transform the system of equations [8] into the differential equation of the second order

$$\left(\partial_{\xi\xi} - \frac{1}{2g_{\xi\xi}}\partial_{\xi}g_{\xi\xi} + \frac{\tilde{j}}{2}\sqrt{\frac{g_{\xi\xi}}{g_{\varphi\varphi}^3}}\partial_{\xi}g_{\varphi\varphi} - \tilde{j}2\frac{g_{\xi\xi}}{g_{\varphi\varphi}} + E^2 g_{\xi\xi}\right)u_j = 0. \tag{13}$$

One can suppose cylindrical geometry in order to simplify the equation into

$$\left(\partial_{\xi\xi} + E^2 - \frac{\tilde{j}2}{R^2}\right)u_j = 0, \tag{14}$$

if a radius vector of the point at the surface will have the form

$$\vec{R} = (R\cos\varphi, R\sin\varphi, \xi), \tag{15}$$

Figure 2.
The simplest realization of smallest graphitic wormholes.

with R as the radius of the cylinder. The solution of this equation has the form

$$u_j(\xi) = Ae^{k\xi} + Be^{-k\xi}, \quad k = \sqrt{\frac{\tilde{j}2}{R^2} - E^2}. \tag{16}$$

A similar form has the dispersion relation associated with the massive 1D Dirac equation

$$k = \sqrt{M^2 - E^2}, \tag{17}$$

where M is the mass of corresponding fermion. It is proven [9] that for a suitable choice of the parameters, 2D massless case is in analogy with 1D massive case, and one can rewrite Eq. (13) in the form

$$\left(\partial_{\xi\xi} - \frac{1}{2g_{\xi\xi}}\partial_\xi g_{\xi\xi} + \frac{\tilde{j}}{2}\sqrt{\frac{g_{\xi\xi}}{g^3_{\varphi\varphi}}}\partial_\xi g_{\varphi\varphi} - \tilde{j}2\frac{g_{\xi\xi}}{g_{\varphi\varphi}} + (E^2 - M^2)g_{\xi\xi}\right)u_j = 0, \tag{18}$$

where M is the mass of the corresponding fermion. For different values of M, one can find the corrections of local density of states for the graphitic wormhole. It seems that these massive particles arising in the wormhole nanotubes could create energy bulks on and near the wormhole bridge which should be experimentally measured by the STM or Raman spectroscopy [10]. Another possibly identification of wormhole structure comes from the fact of creation of strain solitons and topological defects by massive particles on the bridge of bilayer graphene which should propagate throughout the graphene sheet. These are almost macroscopic effects and should be caught by the experimental physicists [11].

1.3 Spin-orbit coupling in the wormhole connecting nanotube

An important measurable quantity in carbon nanostructures, which includes a nanotubular part of a graphitic wormhole, is a spin-orbit coupling (SOC) [12, 13]. If one considers this influence, two-component Dirac equation could be changed into the usual four-component form, and as a consequence chiral fermions should be detected close to the wormhole bridge.

One can reflect on two sources of SOC: (1) the interatomic one that preserves the z-component of a spin and (2) so-called Rashba-type coming from the external electric field, which conserves the z-component of an angular momentum J_z. In both cases, the strength of SOC is influenced by the nonzero curvature. In the next we will be interested in the first source of the SOC.

Considering the SOC we can write the Dirac equation for the nanotube in the form

$$\hat{H}\begin{pmatrix} F_A^K \\ F_B^K \end{pmatrix} = \begin{pmatrix} 0 & \hat{f} \\ \hat{f}^\dagger & 0 \end{pmatrix}\begin{pmatrix} F_A^K \\ F_B^K \end{pmatrix} = E\begin{pmatrix} F_A^K \\ F_B^K \end{pmatrix}, \tag{19}$$

where

$$F_A^K = \begin{pmatrix} F_{A,\uparrow}^K \\ F_{A,\downarrow}^K \end{pmatrix}, \quad F_B^K = \begin{pmatrix} F_{B,\uparrow}^K \\ F_{B,\downarrow}^K \end{pmatrix}. \tag{20}$$

The expression $\hat{f}$ has the following form

$$\hat{f} = \gamma\left(\hat{k}_x - i\hat{k}_y\right) + i\frac{\delta\gamma'}{4R}\hat{\sigma}_x\left(\vec{r}\right) - \frac{2\delta\gamma p}{R}\hat{\sigma}_y, \tag{21}$$

where

$$\hat{k}_x = -i\frac{\partial}{R\partial\theta}, \quad \hat{k}_y = -i\frac{\partial}{\partial y}, \quad \hat{\sigma}_x\left(\vec{r}\right) = \hat{\sigma}_x\cos\theta - \hat{\sigma}_z\sin\theta. \tag{22}$$

Next one can take

$$\gamma = -\frac{\sqrt{3}}{2}aV^{\pi}_{pp}, \quad \gamma' = -\frac{\sqrt{3}}{2}a\left(V^{\sigma}_{pp} - V^{\pi}_{pp}\right), \quad p = 1 - \frac{3\gamma'}{8\gamma}, \tag{23}$$

where a is the length of the atomic bond and V^{σ}_{pp}, V^{π}_{pp} are the hopping integrals for the σ and π bond, respectively.

For the interatomic source of the SOC, one has

$$\delta = \frac{\Delta}{3\varepsilon_{\pi\sigma}}, \quad \Delta = i\frac{3\hbar}{4m^2c^2}\left\langle x_l|\frac{\partial V}{\partial x}\hat{p}_y - \frac{\partial V}{\partial y}\hat{p}_x|y_l\right\rangle \tag{24}$$

with the difference of energies of the relevant π and σ orbitals

$$\epsilon_{\pi\sigma} = \epsilon^{\pi}_{2p} - \epsilon^{\sigma}_{2p}, \tag{25}$$

x_l, and y_l being the local coordinates. By applying the transformation

$$\hat{H}' = \hat{U}\hat{H}\hat{U}^{-1}, \qquad \hat{U} = \begin{pmatrix} \exp\left(i\hat{\sigma}_y\frac{\theta}{2}\right) & 0 \\ 0 & \exp\left(i\hat{\sigma}_y\frac{\theta}{2}\right) \end{pmatrix} \tag{26}$$

the transformed Hamiltonian $\hat{H}'$ will have the form with two terms, including the $\hat{H}_{SOC}$ term which corresponds to the spin-orbit coupling

$$\hat{H}' = \hat{H}_{kin} + \hat{H}_{SOC}, \hat{H}_{kin} = -i\gamma\left(\partial_y \mathrm{Id}_2 \otimes \hat{s}_y + \frac{1}{R}\partial_\theta \mathrm{Id}_2 \otimes \hat{s}_x\right), \hat{H}_{SOC} = \lambda_y\hat{\sigma}_x \otimes \hat{s}_y - \lambda_x\hat{\sigma}_y \otimes \hat{s}_x. \tag{27}$$

The operators $\hat{s}_{x,y,z}$ are the Pauli matrices, which transform the wave function of the A sublattice into the wave function of the B sublattice and vice versa.

In our model, the SOC is induced by the curvature, and it is described with the help of two strength parameters, namely, λ_x and λ_y in the form

$$\lambda_x = \frac{\gamma}{R}\left(\frac{1}{2} + 2\delta p\right), \quad \lambda_y = -\frac{\delta\gamma'}{4R}, \tag{28}$$

for the case of single-wall carbon nanotube with different magnitude. Here, $|\lambda_y| \ll |\lambda_x|$ and for $R \to 0$, both strengths go to infinity. So reminding the previous results, the chiral massive fermions should be detected around the wormhole bridge. For more complicated forms as perturbed nanotube in the wormhole center, the geometry of the corresponding graphene sheets will be curved, and this brings a significant change of the physical properties.

1.4 Graphene black hole

The effects connected with the deformation of a graphene and a consequent change of the distance of the carbon atoms in the layer are described in [14]. It causes the rotation of the p_z orbitals and rehybridization of the π and σ orbitals. The procedure leads to the creation of the $p-n$ junctions similarly to the case of a transistor. This effect changes the Fermi level which is rising in the far areas from the wormhole center. The electron flux is directed from these areas to the middle where the electric charge is accumulated, and in the case of the deformed wormhole, one can speak about so-called graphene black hole. The form of a middle part of the nanotube plays a big role for this purpose. It cannot be unperturbed because in such a case the effect of the black hole would be disrupted. It can be ensured only in the case when the nanotubular neck is tapering in the direction to its center, because this ensures the decrease of the Fermi level [15]. The related effects which appear on the nanostructures are also described in [16], where the special relativistic-like properties of the Beltrami pseudosphere naturally point to quantum field theory in curved space. In the work the finite temperature local density of states is predicted that is a realization of the Hawking-Unruh effect. Mentioned effect of the graphene black hole could eventually disappear in the presence of external magnetic (electric) field which would cause the transfer of the charge from one wormhole sheet to another one through a nanotube center. This serves as an important model for further investigations of the electron flux in the presence of the defects with the applications in cosmological models.

2. Spinor fields in biological systems

One of the present problems in genetic engineering is the prediction of biological gene variation and the representation of corresponding genetic code. This issue emerges in the plotting graphs related to the connection curvature of a docking processes. The docking process is important in the genes of the protein structure and could be adopted instead of using a very long alphabet notation as the string sequence and the comparison of the sequences of docking. From this point of view, methods of quantum field theory, general relativity, and related tools can be of high interest. The equilibrium between the supersymmetry and the mirror symmetry of the left-handed and right-handed DNA, RNA, nucleic and amino acid molecules can be explained by anti-de Sitter (AdS) correspondence in the Yang-Mills theory and the Chern-Simon currents in biology as the curvature of the spectrum in genetic code of the protein curvature.

Today, a genetical structure is studied by standard alphabet codes A, T, C, G, and U as a sequence of strings for the representation of genetic code for various organisms without any exact definition of a new time series of genetic code [17] in contrast to standard time series modeling. With this representation [18], it is very difficult to calculate the genetic variation [19] and to perform calculations within a framework of self-consistent mathematical theory [20], namely, in the context of string theory and M- and G-theories [21, 22].

There are still attempts to perform empirical data analysis of the genetic variation [23] and to detect the pattern matching over the gene sequence by using algorithm over a standard alphabet code as their time series representation. It seems that one main problem in this field is how can we predict the genetic variation and the gene structure in the viral particle and other organisms, or in the context of new representation, the question is how we can explain the intuition behind a definition of new time series data of gene, e.g., involved in the Batalin-Vilkovisky cohomology of DNA and the viral gene structure. The Chern-Simons current and the anomaly over a superspace of cell membrane can be applied to diagnose new gene diseases, the cloning technology or the gene therapy in medicine. Moreover, a presented method can be improved also in view of describing a useless trash area of DNA, which is considered as unknown part of human genome.

On the other side, another approach based on the usage of a spinor field in the Kolmogorov space of the time series data [24] over the genetic code can represent the gene structure as the ghost and anti-ghost fields of the codon and anticodon. This can be achieved in the frameworks of supersymmetry [25] and G-theory [22]. Results of the works show that all calculations over the codon can be assumed as a new superspace of the time series representation of the gene structure [24].

In [26–28], we introduced a new representation of the genetic code in the time series using a modeling by strings and D-branes. By applying a spinor field to a superspace in time series data [29], the method allows us to develop supersymmetry for living organisms. In particular, it is possible to control the anomalies in the codon and anticodon ghost fields and construct an algebraic approach for the trash DNA.

The "gravitational" analogy of the Chern-Simons currents in a gravitational physics, emanating from a system of DNA-RNA transcriptions, could have interesting counterparts also in biology. A representation of codons in human genome, derived from the Chern-Simons currents, can be useful in biology to explain the source of connections over protein-docking states. In this perspective, adopting cohomology in biology can be useful as a new modeling tool for plotting genes with spinor field in time series data. Especially, the junk area of DNA, with repeated inactive genes, can be represented by the Chern-Simons currents with extended structures of knot states in a Laurent polynomial of knots.

Further we discuss the role of spinors in the time series of genetic code. We can denote $\mathbb{H}P^1$ as a quaternionic projective space and $H_0(X_t)$ as a pointed space of DNA alphabet sequence with $[A]$, $[T]$, $[C]$, and $[G]$ as an equivalent class of $[A], [T], [C], [G] \in H_0(X_{t,DNA}) := \Phi_i(X_t)$ a ghost field with parity two with $H_0(X_t) = H_0(X_t, *)$ where $* = \{*\}$ is a pointed space. We define an equivalent class of DNA–RNA translation processes by using the notation of a master equation for an interaction between the viral RNA and the host cell DNA by $\{x_t, y_t\} = \{DNA, RNA\}$. The whole state space model of the viral replication cycle, embedded in the host cell, is denoted by $X_{t,DNA}/Y_{t,RNA} = Z_{t,GENE} = \mathbb{H}/\mathbb{H} = \mathbb{H}P^1$ as a moduli state space model with the definition of genetic code as an equivalent class of the map $\alpha_t : X_{t,DNA} \to \{[A],[T],[C],[G]\} \subset \mathbb{H}$; the host cell gene alphabet is defined by a hidden state space X_t with the gene β_i

$$\begin{aligned}
[A]_{DNA} &:= \left[e^{\frac{i\pi\beta_i}{2}}\right] + [0]\mathbf{i} + [0]\mathbf{j} + [0]\mathbf{k}, \\
[T]_{DNA} &:= [0] + \left[e^{\frac{-i\pi\beta_i}{2}}\right]\mathbf{i} + [0]\mathbf{j} + [0]\mathbf{k}, \\
[C]_{DNA} &:= [0] + [0]\mathbf{i} + \left[e^{i\pi\beta_i}\right]\mathbf{j} + [0]\mathbf{k}, \\
[G]_{DNA} &:= [0] + [0]\mathbf{i} + [0]\mathbf{j} + \left[e^{i2\pi\beta_i}\right]\mathbf{k}.
\end{aligned} \tag{29}$$

In the retroviral RNA of the observed state space $Y_{t,RNA}$ is a span by gene α_i with the anti-ghost field $\Phi_{i,+}(Y_{t,RNA})$ of viral particle. We define a pair of ghost and anti-ghost field genes by a middle hidden state in mRNA and ribosomal EPA state in codon and anticodon state as the ghost and anti-ghost fields in the genetic code. One can define a mutual genetic code as passive or dual hidden states $[s_1]^*, \ldots, [s_8]^*$ and active eight states $[s_1], \ldots, [s_8]$ for the spinor field in the genetic code by

$$\begin{aligned}
[A]_{tRNA} &:= [NU]_{mRNA} = \left[e^{\frac{j\pi\alpha_i}{2}}\right] + [0]\mathbf{i} + [0]\mathbf{j} + [0]\mathbf{k}, \\
[U]_{tRNA} &:= [NA]_{mRNA} := [0] + \left[e^{\frac{-j\pi\alpha_i}{2}}\right]\mathbf{i} + [0]\mathbf{j} + [0]\mathbf{k}, \\
[C]_{tRNA} &:= [NG]_{mRNA} := [0] + [0]\mathbf{i} + \left[e^{j\pi\alpha_i}\right]\mathbf{j} + [0]\mathbf{k}, \\
[G]_{tRNA} &:= [NC]_{mRNA} := [0] + [0]\mathbf{i} + [0]\mathbf{j} + \left[e^{j2\pi\alpha_i}\right]\mathbf{k}.
\end{aligned} \tag{30}$$

The reversed transcription process of the gene expression is defined by a moduli state space model of a coupling spinor field between the gene of a viral particle and the host cell (**Figure 3**)

$$\mathbb{H}P^1 = X_t/Y_t \ni \left[1, \frac{e^{\frac{2\pi}{4}ni\alpha}}{e^{\frac{2\pi}{4}mj\beta}}\right]_{m,n=1,2,3,4} = \left[1, \frac{q|_{DNA}}{q^*|_{RNA}}\right] = \left[\frac{q|_{DNA}}{q^*|_{RNA}}, 1\right]. \tag{31}$$

One can define $Sp(1) \to S^7 \to \mathbb{H}P^1$ as a Hopf fibration of eight states of the genetic code $[s_1], [s_2], \ldots [s_8] \in S^7 = T_p\mathcal{M}$, denoted by $[s_1]^*, [s_2]^*, \ldots [s_8]^* \in T_p^*\mathcal{M}$ states of the genetic code of the space of a viral RNA X_t and a space of host cell DNA, Y_t.

If $\mathcal{U}_{[A]_\alpha} \subset \mathbb{H}P^1$ is a chart of local coordinate in a manifold of genetic code over X_t/Y_t, where $[A]_\alpha$ is defined over the right-hand isomer genetic code $\{A, T, C, G\}$ (for the simplicity we use a symbol G also for U) with their dual $[A]_\alpha^*$, with the mirror symmetry of a genetic code $\{NA, NT, NC, NG\}$. We have a cycle and a cocycle of an orbifold as a trivialization over the tangent of the living organism manifold, so-called codon and anticodon $\mathcal{U}_i \cap \mathcal{U}_j \cap \mathcal{U}_k$. Let $(\mathcal{M}, g)$ be a living organism manifold with $\mathcal{M} = \mathbb{H}P^1$ for a living organism with the Riemannian metric tensor $g_{ij} = <T_{[A]_{[A]_\alpha}}\mathcal{M}, T^*_{[A]_\alpha}\mathcal{M}>$ over a tangent manifold and a cotangent manifold

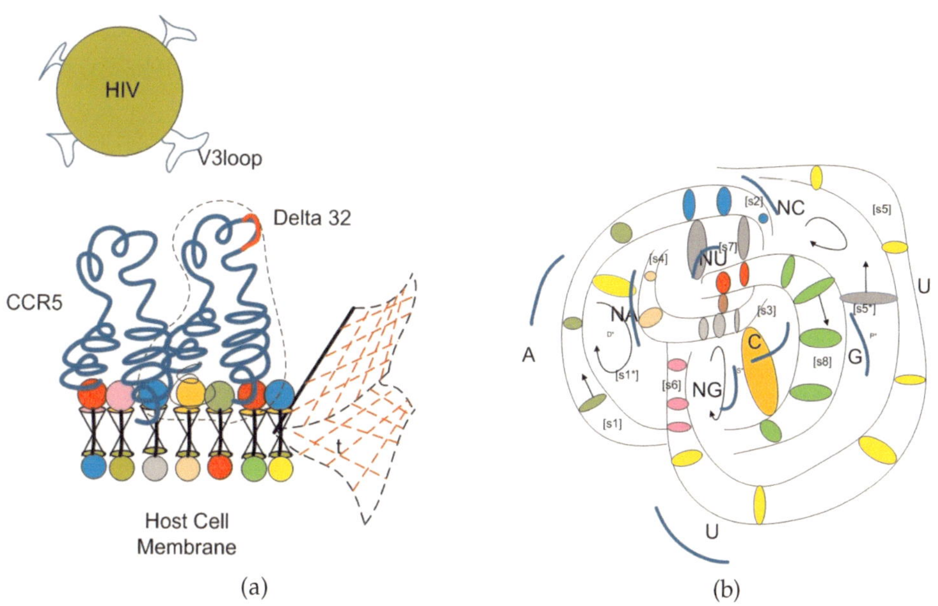

Figure 3.
(a) The site of docking between CCR5 Δ32 co-receptor in host T-cell membrane and V3 loop in HIV viral gp120. The cell membrane has a mirror symmetry structure of D-brane for outer layer. The anti-D-brane structure of the cell membrane is an inner layer of phospholipid. (b) The knot 4_1 model of a short exact sequence of hidden eight states in genetic code.

$T_{[A]_\alpha}\mathcal{M} = S^7$ with g_{ij} defined as a tensor behavior field transformation between 64 states in a codon as the distance in a space of the genetic code with $i, j = 1, 2, 3, ...8$. The active states in genetic code are denoted as $[s_1], [s_2], \cdots [s_8]$, and the hidden states inside eight states or passive states are denoted as $[s_1^*], [s_1^*], \cdots [s_8^*]$. The dual part of a genotype $[s_i^*]$ is defined by a tangent of a manifold of the genetic code with the Jacobian $J = \sqrt{g_{ij}}$, where the metrics is $g_{ij} = < [s_i, [s_j]^* >$. A smallest state in gene is defined as a pair of a genetic code by a classical notation $A - T$, $C - G$ with the coordinates (A, T), (C, G). One can define a superstate in a pair of genes as ghost and anti-ghost fields in the genetic code with the supersymmetry of D-isomer to L-isomer from the right-hand D-state in the light of the polarization in a nucleic acid as $[s_i]|_{\Phi_i}$ to a left-hand light L-state of an isomer of a light polarization, denoted as $[s_i]|^*_{\Phi_i^+}$

$$\begin{aligned}
&[s_1] = ([A], [T]^*) \in T_p\mathcal{M}, && [p1] = [s_1]^* = [s_{11}]^* = ([A], [T]^*)^* \in T_p^*\mathcal{M}\\
&[s_2] = ([A], [NA]) \in T_p\mathcal{M}, && [s_2]^* = ([A], [NA])^* \in T_p^*\mathcal{M}\\
&[s_3] = ([C], [G]^*) \in T_p\mathcal{M}, && [s_3]^* = ([C], [G]^*)^* \in T_p^*\mathcal{M}\\
&[s_4] = ([C], [NC]) \in T_p\mathcal{M}, && [s_4]^* = ([C], [NC])^* \in T_p^*\mathcal{M}\\
&[s_5] = ([T], [T]^*) \in T_p\mathcal{M}, && [s_5]^* = ([T], [T]^*)^* \in T_p^*\mathcal{M}\\
&[s_6] = ([T], [NA]) \in T_p\mathcal{M}, && [s_6]^* = ([T], [NA])^* \in T_p^*\mathcal{M}\\
&[s_7] = ([G], [G]^*) \in T_p\mathcal{M}, && [s_7]^* = ([G], [G]^*)^* \in T_p\mathcal{M}\\
&[s_8] = ([G], [NC]) \in T_p\mathcal{M}, && [s_8]^* = ([G], [NC])^* \in T_p\mathcal{M}
\end{aligned} \tag{32}$$

As known from biochemistry, there exist only $[s_1] = ([A], [T]^*)$ and $[s_3] = ([C], [G]^*)$ observational states of the living organism in nature. However, in the theory of supersymmetry of S^7 Hopf fibration, there exist eight states of the ghost fields with six hidden states in the mirror symmetry. In each state, an orbifold of the living organism, other eight states denoted p_1–p_8 exist. One can define all 64 states with the help of a notation for the Riemann tensor field $g_{ij} = < [s_i], [s_j]^* >$, e.g., $g_{11} = < [s_1], [s_1]^* >$, and we denote a pair of states in a gene as $([s_1], [p_1])$. It is a pair of the genetic code $g_{11} = < ([A], [T]^*), ([A], [T^*]) >$; for the case of codon, we replace $[T]$ with $[U]$. Finally, we have for $g_{11} = < ([A], [[[T]^*)]], ([A], [T^*]) > = < ([A], ([A], [T^*]) >$ since the states $[T^*]$ are hidden states, so one has a codon AAU for g_{11}.

If $\Gamma_{ij}^k := [A]_k$ is a connection over a tangent of a manifold of X_t/Y_t with $\mathcal{M} = \mathbb{HP}^1$ of the genetic code $k \in \{A, T, C, G\}$, one can denote $\Gamma_{\mu\nu}^k := F_{\mu\nu}$ as the behavioral Yang-Mills field with its dual in AdS theory of supersymmetry with $*F_{\mu\nu} = F^{\mu\nu}$. It is a behavior of a protein folding inducing a curvature between a viral glycoprotein and a host cell receptor. The behavior of a Yang-Mills field is an interaction field between the behavior of virus and host cells which can survive by a change of curvature of the protein during the evolution. It is a connection in the sense of an evolutional field not in the sense of traditional gravitational field as usual

$$\Gamma_{ij}^k = \frac{1}{2} g^{kl} \left(\partial_j g_{jk} + \partial_i g_{jk} - \partial_k g_{jk} \right), \tag{33}$$

where $g_{ij} = < [s_i], [s_j]^* >$.

For an equilibrium state of evolution of an organism, we have no change of covariant derivative for the tensor field g

$$\nabla_g g = 0. \tag{34}$$

We can define a $F^{\mu\nu} := (A_\alpha)^\nu_\mu$ as a connection on a principle bundle $Sp(1) \to S^7 \to \mathbb{H}P^1$ of a genetic code over a supermanifold of a viral particle $\mathcal{A}$. We can use an exact sequence of the sheave cohomology $\mathcal{O}_{X_t}$ with the chart over the supermanifold defined by homogeneous coordinates of $\mathbb{H}P^1$ for viral gene X_t along the host cell gene $\mathcal{O}_{Y_t}$, while the virus attachment to the host cell is defined by the coordinate in sheave $\mathcal{O}_{X_t/Y_t}$. We apply a supersymmetry AdS theory over a Yang-Mills field of a behavioral field of the genetic code as a connection over $(A_\alpha) \in \{[A], [T], [C], [G]\} \subset \mathbb{H}P^1$ with the Hopf fibration of viral RNA gene $F^{\mu\nu} := (A_\alpha)^\mu_\nu = \Gamma^\mu_{\alpha\nu}$ with an anti self dual field over the gene of the host cell DNA, $*F_{\mu\nu}$. A current J of the connection between the fields is defined by the Chern-Pontryagin density for the interaction of behavioral fields. The current varies from the curvature of docking between the behavior of the curvature over amino acid k in X_t and its dual curvature in Y_t while docking $<F^{\mu\nu} * F_{\mu\nu}>$, where $< - >$ is an average or an expectation operator.

The connection between genes is

$$R^\mu_{\nu\alpha\beta} = \partial_\alpha \left(A_\beta\right)^\mu_\nu - \partial_\beta (A_\alpha)^\mu_\nu + \left[A_\alpha, A_\beta\right]^\mu_\nu, \tag{35}$$

where $R^\mu_{\nu\alpha\beta}$ is the curvature over a tangent space of a genetic code $(A_\alpha) \in \{[A], [T], [C], [G]\} \subset \mathbb{H}P^1$ and $(A_\alpha)^\mu_\nu := \Gamma^\mu_{\alpha\nu} = F^{\mu\nu}$ is the connection of the coupling between two alphabets of two organisms, i.e., from DNA and viral RNA. The above curvature is also useful in other situations like t-RNA docking with DNA in interaction between two D-branes of DNA and RNA. In a gauge field theory of DNA and RNA genetic codes of translation process, it is the group action of Lie-algebra one form. We have an adjoint representation of the genetic code as a translation process over the codon and anticodon of $t - RNA$. One needs to define a new value to measure the curvature in amino acids of protein structure, not in the tangent space of a genetic code, and one needs to define a curvature over the codon and anticodon to represent a curvature of proteins while they are docking. The new value needs to be unique for all states in the codon and to have a meaning of curvature of genetic code with connection over manifold of genetic code. To represent a spectrum of genetic variation as curvature in protein structure while docking, one can introduce a new quantity, so-called Chern-Simons current for biology.

One can translate a genetic code in a codon in three steps. The translation operator of group is given by a behavior matrix in Lie group, a group of supermanifold of living organism with action in three times. It generates a codon representation as an adjoint representation over gene expression, and it is a precise definition of genetic code with parity two of ghost field and anti-ghost field in the Chern-Simons current for the representation of a gene A_i with the current density $J^{A_i} = \int_{t_1}^{t^n} dJ^{A_i}$.

Let a knot serve as a representation of anticodon in t-RNA topological structure for amino acid μ with J^μ as a representation R of gauge group G of gene geometric translation as group action of transcription process; then the genetic code is an average expectation value of Wilson loop operator of coupling between hidden state of x_t and y_t twist D-brane and anti-D-brane over superspace of cell membrane, i.e.,

$$W(K,R) = \mathrm{Tr}_R P \exp\left(\oint_K A\right). \tag{36}$$

The above term gives the asymmetric property of chiral molecule of DNA and RNA, twisted from the left hand to the right hand in a supersymmetry breaking as knot polynomial related to the connection A. By such a way, we can represent the genetic code as Laurent polynomials in variable q with integer coefficients, and for any knot K we have

$$J(K,q) = \sum_{i=1}^{n} a_n q^n. \tag{37}$$

By using the new parameter of knot q

$$q = e^{\frac{2\pi}{k+h}} \tag{38}$$

one can induce a spinor field for representation of genetic code, where h is the dual coexter number for group action of supersymmetry of gene expression G. It might be the source of evolution from the adaptive behavior derived from the environment. In the next, we set $h = 0$ in our definition of the Chern-Simons current for biology for the simpler derivation of formulas.

A Chern-Simons current J^μ for anomaly quantum system of codons can be also defined as the spectrum of curvature in the genetic code for gene evolution detection. Under the definition we mean a differential 3-forms in cohomology of spin fiber S^3 over the homotopy class $\left[S^3, X_t/Y_t\right]$ in the codon of t-RNA molecule. A path integral of gene expression is defined by the Chern-Simons theory over knots of codon and anticodon: it is defined by the interaction between codon A_i and anticodon between DNA and RNA in the form of integral $A_i + S_{CS} = \int \mathcal{D}A_i \exp\left(iS_{CS}\right)$

$$S_{CS} = \frac{k}{4\pi}\int_W \mathrm{Tr}\left(A \wedge dA + \frac{2}{3} A \wedge A \wedge A\right) \tag{39}$$

and

$$\begin{aligned} J\left(q; K^{A_i}, R_i\right) &= <W(K_i, R_i)> \\ &= <\mathrm{Tr}_{R_i} P \oint_{K_i} A> = \frac{\int \mathcal{D}A_i \exp\left(iS_{CS}\right) \Pi_i W(K_i, R_i)}{\int \mathcal{D}A_i \exp\left(iS_{CS}\right)}. \end{aligned} \tag{40}$$

The explicit definition of curvature over the connection of genetic code has also new meaning of the genetic spectrum current J^μ, $\mu = 1, 2 \cdots 20$ of the Chern-Simons current; it is generated from the representation of Lie group over manifold of a host cell.

An example of our approach can serve as a case of phenylalanine (*Phe*), where UUU and UUC definition is

$$\begin{aligned} J^{Phe} &= \epsilon^{\mu\alpha\beta\alpha} < \frac{1}{2} A_\alpha \partial_\beta A_\gamma + \frac{1}{3} A_\alpha A_\beta A_\gamma > \\ &= \epsilon^{\mu\alpha\beta\alpha} < \frac{1}{2} (A_2)^\mu_\nu d(A_2)^\mu_\nu + \frac{1}{3} (A_2)^\mu_\nu (A_2)^\mu_\nu (A_2)^\mu_\nu > \end{aligned} \tag{41}$$

where we explicitly define the differential form of genetic code for *Phe* by $dA_2 = A_2A_2 - A_2A_3 := UU - UC$, so we have

$AdA = A_2(A_2A_2 - A_2A_3) = A_2A_2A_2 - A_2A_2A_3$. The minus sign represents a linear combination of basis for codon. Therefore it follows

$$\begin{aligned} J^{Phe} &= \epsilon^{\mu\alpha\beta\alpha} < \frac{1}{2}A_\alpha\partial_\beta A_\gamma + \frac{1}{3}A_\alpha A_\beta A_\gamma > \\ &= \int \mathrm{Tr}\left(\frac{5}{6}U \wedge U \wedge U - \frac{1}{2}U \wedge U \wedge C\right) = \int \mathrm{Tr}\left(H^3(\mathcal{M})\right). \end{aligned} \tag{42}$$

For a translation in reversed direction of antigen shift and drift in gene evolution theory, we can use the definition of the group action of reversed direction of time by the *CPT* theory for anti-ghost field in field of time series of antibody gene as

$$\begin{aligned} \{x_t, y_t\} &= \int \mathrm{Tr}\left(H^3(\mathcal{M})\right) = \sum_{i=1}^{3} g^i\left(x_t/y_t\right) \\ &= \alpha_t y_t, g^3 x_t/y_t \rightarrow \beta_t x_t/\alpha_t y_t \simeq \left[\epsilon_t^*\right] = \int \mathrm{Tr}H^3\left(\mathcal{M}, g, F^{\mu\nu}\right). \end{aligned} \tag{43}$$

Then a numerical representation for spinor field of curvature in the gene expression by the Chern-Simons action is defined as follows:

$$S_{CS} = \frac{k}{4\pi}\int \mathrm{Tr}\left(A \wedge dA + \frac{2}{3}A \wedge A \wedge A\right) \tag{44}$$

where $k = 1, 2, 3 \ldots n$ are

$$J^{\mathrm{Amino}} = J^k \simeq \sqrt{\frac{2}{k+2}}\sin\frac{\pi}{k+2} = \int D[A]^{S_{CS}}. \tag{45}$$

The derivation of the Chern-Simons current can be done by a simple algorithm [30], i.e., the Chern-Simons current maps the string of genetic code into numerical values by explicit formulas. It can be used to plot the time series data directly into the superspace of gene expression. We transform the alphabet string values, which cannot be computed in the classical standard definition of genetic code, into the Chern-Simons current of time series data of genetic code with $k = 1, 2, 3, \ldots 64$ over spinor field with ground field of real values. We think that the approach is more suitable for computational programs used in data analyses.

3. Circular Artin braid group representation for spinor field in genetic code

In each cell division, the telomeres are shortened [31], and total length of DNA is changing. As the result of shorter biological clock from cell division, the living things die. In order to understand cell cocycle and division mechanism of telomerase aging, one can explain the source of cancer as a source of age acceleration and its relationship to telomere shortening mechanism. It is a source of braid group operation [32] so-called self-diffeomorphism in the genetic code. The age acceleration is a relative measurement between the chronical clock and the biological clock in telomere. Up to now, scientists understand that a telomere and telomerase are the locations of ancient viruses that rely on DNA in the chromosomes of living organisms. Telomere is composed of the repeated sequence of $(TTAGGG)_{dt^*}$ where

$1000 \le dt^* \le 2000$. The size of the duplicate sequence at the end of this open chromosome is amplified by six braids caused by six superspaces in time series data of organisms. The G alphabet might be suitable to be chosen as hidden time scale in the biological clock.

Here we assume that all genetic code cannot be completely separated and biological clock in telomere length is parametrized by a hidden state of the number of $dt^* := [G]$ alphabet in $(TTAGGG)_{n=dt^*}$ repeated pattern of the telomere. The element is Grothendieck topology over an adjoint cofunctor; it is a self-diffeomorphism $\xi : B_3^c \to B_3^c$. The loop braid generator for B_3^c is a quaternionic field in genetic code. We define their explicit forms and their permutations over the symmetric group by a chosen basis in Clifford algebra as

$$\sigma_D \to \left[\frac{1}{2}, \frac{1}{2}, -\frac{1}{2}\right], \quad \sigma_R \to \left[-\frac{1}{2}, \frac{1}{2}, \frac{1}{2}\right], \quad \sigma_P \to \left[\frac{1}{2}, -\frac{1}{2}, \frac{1}{2}\right], \tag{46}$$

and $\sigma_{D^*} = \sigma_D^{-1}$, $\sigma_{R^*} = \sigma_R^{-1}$, and $\sigma_{P^*} = \sigma_P^{-1}$. We have

$$\sigma_{D^*} \to \left[-\frac{1}{2}, -\frac{1}{2}, \frac{1}{2}\right], \quad \sigma_{R^*} \to \left[\frac{1}{2}, -\frac{1}{2}, -\frac{1}{2}\right], \quad \sigma_{P^*} \to \left[-\frac{1}{2}, \frac{1}{2}, -\frac{1}{2}\right], \tag{47}$$

therefore one can write eight bases for spinor field in the genetic code in braid form as follows

$$\begin{aligned}
\sigma^{[G]} &:= [0,0,0]\sigma_R^{-1}\sigma_D = \sigma_R^{-1}\sigma_D,\\
\sigma^{[A]} &:= [0,0,1]\sigma_R^{-1}\sigma_D = (\sigma_D\sigma_R)(\sigma_R^{-1}\sigma_D),\\
\sigma^{[U]} &:= [0,1,0]\sigma_R^{-1}\sigma_D = (\sigma_P\sigma_D)(\sigma_R^{-1}\sigma_D),\\
\sigma^{[C]} &:= [1,0,0]\sigma_R^{-1}\sigma_D = (\sigma_R\sigma_P)(\sigma_R^{-1}\sigma_D),\\
\sigma^{[NA]} &:= [0,1,1]\sigma_R^{-1}\sigma_D = (\sigma_R\sigma_P)(\sigma_P\sigma_D)(\sigma_R^{-1}\sigma_D),\\
\sigma^{[NU]} &:= [1,1,0]\sigma_R^{-1}\sigma_D = (\sigma_D\sigma_R)(\sigma_R\sigma_P)\sigma_R^{-1}\sigma_D,\\
\sigma^{[NC]} &:= [1,0,1]\sigma_R^{-1}\sigma_D = (\sigma_P\sigma_D)(\sigma_D\sigma_R)(\sigma_R^{-1}\sigma_D),\\
\sigma^{[NG]} &:= [1,1,1]\sigma_R^{-1}\sigma_D = (\sigma_R\sigma_P)(\sigma_P\sigma_D)(\sigma_D\sigma_R)(\sigma_R^{-1}\sigma_D).
\end{aligned} \tag{48}$$

We may also use $\theta = 2\pi s$ with spin quantum number s being an integer for retrotransposon and half-integer for geneon, so that

$$e^{i\theta} = e^{2i\pi s} = (-1)^{2s}\psi_2\psi_1. \tag{49}$$

In a 3-dimensional position space, the geneon and retrotransposon statistics operators are -1 and $+1$, respectively. By the same way, in two-dimensional position space, the abelian anyonic statistics operators $e^{i\theta}$ are 1-dimensional representations of eight loop braid elements $\sigma_1, \sigma_1^2, \sigma_1^3, \cdots, \sigma_1^8$ in circular Artin braid group B_2^3 acting on the space of wave functions (**Figure 4**).

3.1 Classification of loop braid group in genetic code

We classify three types of loop braid group operations; it is a representation of an anyon for protein folding. For two-dimensional representation of D-brane in loop braid group for the genetic code, we define abelian anyon for biology in (2 + 1) dimensions, the extra dimensions used to represent the homotopy path of protein folding.

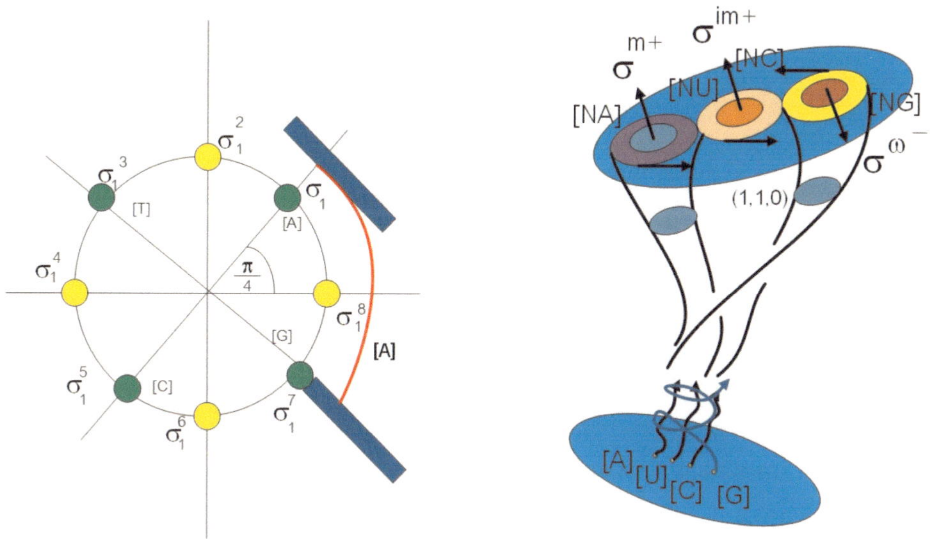

Figure 4.
The left picture shows biological Artin braid element $\sigma_1 \in B_2$ in complex plane $\mathbb{C}$. There are eight equivalent classes span by eight orders of σ_1. It is a representation of eight states in three genetic codes in codon as braid element in $\left(\sigma_1^{[A_\mu]}, \sigma_1^{[A_\mu]}, \sigma_1^{[A_\mu]}\right) \in B_2^3$. The red color line represents the curvature from the physiology of biological time series data, and the blue color represents the active and passive behavior field layers. The right picture shows a member of loop braid group. Three circles S^1 represent the sources of closed 3-balls B_3^c, the structure group of affine transform of 3 behavior fields in genetic code $\sigma_i^{[A_\mu]} \in B_2^3$ action on affine fibre bundle of behavior field in the genetic code $\sigma^{\omega^\pm}$, $\sigma^{im^\pm}$, $\sigma^{m^\pm} \in B_3^c$. We have affine group as loop braid group in genetic code by $B_2^3 \rtimes B_3^c$.

The loop braid group, LB_n^{BIO}, for a genetic code has three types of generators, $\sigma = (\sigma_D, \sigma_R, \sigma_P)$, $\rho = \left(\sigma^{[A]},, \sigma^{[U]},, \sigma^{[C]},, \sigma^{[G]}\right)$, and $\tau = \left(\sigma^{\omega},, \sigma^{im},, \sigma^{m}\right)$. The generators $\left\{\sigma_i, \rho_i \in B_2^3 i = 1, \cdots, n-1\right\}$ and $\left\{\tau_i \in B_3^c i = 1, \cdots, n\right\}$ fulfill the relations

$$
\begin{aligned}
\tau_i \tau_j &= \tau_j \tau_i, && i \neq j \\
\tau_i^2 &= 1, && i = 1, \cdots, n \\
\sigma_i \tau_j &= \tau_j \sigma_i, && |i-j|>1 \\
\rho_i \tau_j &= \tau_i \rho_j, && |i-j|>1 \\
\tau_i \rho_i &= \rho_i \tau_{i+1}, && i = 1, \cdots, n-1 \\
\tau_i \sigma_i &= \sigma_i \tau_{i+1}, && i = 1, \cdots, n-1 \\
\tau_{i+1} \sigma_i &= \rho_i \sigma_i^{-1} \rho_i \tau_i, && i = 1, \cdots, n-1.
\end{aligned}
\tag{50}
$$

For a group operation of the genotype G and a representation ρ for gene translation as an anyon, we have $\rho : G \to U(1) = S^2$ as a representation in Ω_n^{BIO} and in LB_n^{BIO}. In order to visualize 3D folding structure of the protein structure, we define three types of loop braid group operations in biology. All loop elements of the representation of amino acids arose from group operations over the superspace of time series data. These tree types are a translation, reflection, and rotation, ρ^{Type-I}, $\rho^{Type-II}$, and $\rho^{Type-III}$. For the translation as a string of amino acids, we have the anyon $Type - I$ for biology,

$$
\begin{aligned}
&\rho^{Type-I}(\sigma_i)\rho^{Type-I}(\sigma_{i+1})\rho^{Type-I}(\sigma_i) \\
&\quad = \rho^{Type-I}(\sigma_{i+1})\rho^{Type-I}(\sigma_i)\rho^{Type-I}(\sigma_{i+1}),
\end{aligned}
$$

$$\rho^{Type-I}(\sigma_{i+1})\rho^{Type-I}(\sigma_i)\rho^{Type-I}(\sigma_{i+1}) \\ = \rho^{Type-I}(\sigma_{i+1})\rho^{Type-I}(\sigma_{i+1})\rho^{Type-I}(\sigma_i), \\ \rho^{Type-I}(\sigma_i) = \rho^{Type-I}(\sigma_{i+1}). \tag{51}$$

It is isomorphic to the mapping class group of the infinitely punctured disk, a discrete set of punctures limiting to the boundary of the disk.

By an analogy with the action of the symmetric group by permutations in various mathematical settings, there exists a natural action of the braid group on n-tuples of objects or on the n-folded tensor product that involves some twistors. Let us consider an arbitrary group G, and let X be the set of all n-tuples of elements of G whose product is the identity element of G. The kernel of the homomorphism $LB_n^{BIO} \to \Omega_n^{BIO}$ is a subgroup of LB_n^{BIO} called pure loop braid group for biology on n strands and denoted as LP_n^{BIO}. In the pure braid, the beginning and end of each strand are in the same position. Pure braid groups fit into a short exact sequence

$$1 \to LF_{n-1} \to LP_n^{BIO} \to LP_{n-1}^{BIO} \to 1. \tag{52}$$

This sequence splits, and therefore pure braid groups are realized as iterated semi-direct products of free groups.

The braid group B_3 is the universal central extension of the modular group $PSL(2, \mathbb{Z})$, with these sitting as lattices inside the universal covering group. If we define $\mathcal{O}_D$, $\mathcal{O}_R$, and $\mathcal{O}_P$ as active layers over the superspace of DNA, RNA, and protein, $\mathcal{O}_{D^*}$, $\mathcal{O}_{R^*}$, and $\mathcal{O}_{P^*}$, as passive layers, we can define braid group in genetic code by a curvature inside DNA, RNA, and protein folding structure. It is a source of an acceleration of biological clock in the epigenetic code. We let σ_D, σ_R, σ_P, σ_{D^*}, σ_{R^*}, σ_{P^*}, and $\sigma^{[A_\mu]}$ and $[A_\mu] = [A], [U], [C], [G], [NA], [NU], [NC], [NG]$ be a loop braid group elements in the genetic code. They are the circular Artin braid groups for the genetic code. One can then define

$$\Psi^R = \sigma_D\sigma_R\sigma_D, \quad \Psi^P = \sigma_D\sigma_R, \quad \Psi^{P^*} = \sigma_R\sigma_D\sigma_R. \tag{53}$$

The braid group operation gives Ψ^P

$$\sigma_D\Psi_P{\sigma_D}^{-1} = \sigma_R\Psi^P{\sigma_R}^{-1} = \Psi^P \tag{54}$$

implying that Ψ^P is in the center of B_3. It is a wave function of protein transition anyon state. If $G = LB_n^{BIO}$ acts on X_t, we get

$$\sigma_i^{D^*}\Psi_i(d_1, \ldots, d_{i-1}, d_i, d_{i+1}, \ldots, d_n) \\ = \Psi_i\left(d_1, \ldots, d_{i-1}, d_{i+1}, d_{i+1}^{-1}d_i d_{i+1}, d_{i+2}, \ldots, d_n\right). \tag{55}$$

If $G = LB_n^{BIO}$ acts on Y_t, we get

$$\sigma_i^{R^*}\Psi_i(r_1, \ldots, r_{i-1}, r_i, r_{i+1}, \ldots, r_n) \\ = \Psi_i\left(r_1, \ldots, r_{i-1}, r_{i+1}, r_{i+1}^{-1}r_i r_{i+1}, r_{i+2}, \ldots, r_n\right). \tag{56}$$

If $G = LB_n^{BIO}$ acts on $P_t = X_t/Y_t$, we get

$$\sigma_i^{P^*}\Psi_i(p_1, \ldots, p_{i-1}, p_i, p_{i+1}, \ldots, p_n) \\ = \Psi_i\left(p_1, \ldots, p_{i-1}, p_{i+1}, p_{i+1}^{-1}p_i p_{i+1}, p_{i+2}, \ldots, p_n\right). \tag{57}$$

Thus the elements d_i and d_{i+1} exchange places in DNA strand by an analogy with genetic variation. If d_i is twisted by the inner automorphism corresponding to d_{i+1} at the position $i = dt^*$, the product of the d components remains the identity element. It may be checked that the braid group relations are satisfied and this formula indeed defines a group action of LB_n^{BIO} on X_t.

4. Conclusions

The spin-orbit coupling is an important quantity, which is measurable in the carbon nanostructures, including the graphitic wormhole (or its nanotubular part); it can also help to identify the wormhole structure in details. SOC in a graphene could be induced by the nonzero curvature; in the particular case of the wormhole with negative curvature, the chiral fermions penetrating through the connecting nanotube in the wormhole structure could be created. The two-component Dirac equation is changed into the usual four-component form. As a consequence, the chiral fermions should be detected close to the wormhole bridge; the effect is stronger if the radius of the wormhole bridge is smaller. Moreover, one can detect permanently oriented flow when the chiral fermions prefer only one direction of the massive or massless fermionic current from the upper graphene sheet to the lower one, depending on the wormhole curvature.

We also describe the role of spinor fields in the time series of genetic code. The reversed transcription process of the gene expression could be defined by a moduli state space model of a coupling spinor field between the gene of a viral particle and the host cell. As a general result, all states of codon can be computed by the Chern-Simons 3-forms. The Chern-Simons current, coming from ghost and anti-ghost fields of supersymmetry theory, can be used to define a spectrum of gene expression in new time series data where a spinor field, as alternative representation of a gene, is adopted instead of using the alphabet sequence of standard bases A, T, C, U, and G. Effort is also directed toward the explanation of the adaptive behavior of immunosystem and to find the source of cancer from the physiology of telomere malfunction in DNA repairing state. Similar examination of Holo-Hilbert spectral analysis of the Chern-Simons current in V3 loop genotypes was performed recently in [33]. A genetic variation in V3 loop genotypes was forecast by using the imaging generated from tensor correlation network with an autoregressive integrated moving average model, support spinor model, and convolutional neural network algorithms.

The reported results of the work have promissory perspective for their extension to interdisciplinary areas as machine learning, econophysics, or biological sciences.

Acknowledgements

The work is partly supported by Scientific Grant Agency VEGA Grant No. 2/0009/19 and No. 2/0153/17. R. Pinčák would like to thank the TH division in CERN for hospitality.

Author details

Richard Pinčák[1,2] and Erik Bartoš[3*]

1 Institute of Experimental Physics, Slovak Academy of Sciences, Košice, Slovak Republic

2 Bogoliubov Laboratory of Theoretical Physics, Joint Institute for Nuclear Research, Moscow Region, Russia

3 Institute of Physics, Slovak Academy of Sciences, Bratislava, Slovak Republic

*Address all correspondence to: erik.bartos@savba.sk

References

[1] González J, Herrero J. Graphene wormholes: A condensed matter illustration of dirac fermions in curved space. Nuclear Physics B. 2010;**825**(3): 426-443

[2] Pudlak M, Pincak R. Edge states of graphene bilayer strip. European Physical Journal B. 2013;**86**(3):1-4. Article number: 107

[3] Castro EV, López-Sancho MP, Vozmediano MAH. New type of vacancy-induced localized states in multilayer graphene. Physical Review Letters. 2010;**104**(3):1-4. Article number: 036802

[4] Castro EV, López-Sancho MP, Vozmediano MAH. Vacancy induced zero energy modes in graphene stacks: The case of abc trilayer. Solid State Communications. 2012;**152**(15): 1483-1488

[5] Kochetov EA, Osipov VA, Pincak R. Electronic properties of disclinated flexible membrane beyond the inextensional limit: Application to graphene. Journal of Physics. Condensed Matter. 2010;**22**(39):1-11. Article number: 395502

[6] Rostami H, Asgari R. Electronic ground-state properties of strained graphene. Physical Review B: Condensed Matter and Materials Physics. 2012;**86**(15):1-7. Article number: 155435

[7] Jiang C-W, Zhou X, Xie R-H, Li F-L. Semiclassical molecular dynamics simulations for ultrafast processes in molecules. Quantum Matter. 2013;**2**(6): 353-363

[8] Pincak R, Smotlacha J. Analogies in electronic properties of graphene wormhole and perturbed nanocylinder. European Physical Journal B. 2013;**86**(11):1-7. Article number: 480

[9] Alhaidari AD, Jellal A, Choubabi EB, Bahlouli H. Dynamical mass generation via space compactification. Quantum Matter. 2013;**2**(2):140-143

[10] Tan PH, Han WP, Zhao WJ, Wu ZH, Chang K, Wang H. The shear mode of multilayer graphene. Nature Materials, 11(4):294–300, 2012

[11] Alden JS, Tsen AW, Huang PY, Hovden R, Brown L, Park J, et al. Strain solitons and topological defects in bilayer graphene. Proceedings of the National Academy of Sciences. 2013; **110**(28):11256-11260

[12] Kane CL, Mele EJ. Quantum spin hall effect in graphene. Physical Review Letters. 2005;**95**(22):1-4. Article number: 226801

[13] Ando T. Spin-orbit interaction in carbon nanotubes. Journal of the Physical Society of Japan. 2000;**69**(6): 1757-1763

[14] Atanasov V, Saxena A. Tuning the electronic properties of corrugated graphene: Confinement, curvature, and band-gap opening. Physical Review B: Condensed Matter and Materials Physics. 2010;**81**(20):1-8. Article number: 205409

[15] Pincak R, Pudlak M, Smotlacha J. Electronic Properties of Single and Double Wall Carbon Nanotubes. New York: Nova Science Publishers, Inc; 2012

[16] Iorio A, Lambiase G. The hawking-unruh phenomenon on graphene. Physics Letters, Section B: Nuclear, Elementary Particle and High-Energy Physics. 2012;**716**(2):334-337

[17] José MV, Zamudio GS, Morgado ER. A unified model of the standard genetic code. Royal Society Open Science. 2017; **4**(3):1-13. Article number: 160908

[18] Hoffman M. Straightening out the protein folding puzzle. Science. 1991; **253**(5026):1357-1358

[19] Jagadesh A, Salam AAA, Zadeh VR, Arunkumar G. Genetic analysis of neuraminidase gene of influenza a (h1n1)pdm09 virus circulating in Southwest India from 2009 to 2012. Journal of Medical Virology. 2017;**89**(2): 202-212

[20] Beyl LM. Super-lagrangians. Physical Review D. 1979;**19**(6): 1732-1745

[21] Sepehri A, Pincak R. G-theory and its reduction to f (R)-gravity in FRW universe. International Journal of Geometric Methods in Modern Physics. 2018;**15**(9):1-15. Article number: 1850144

[22] Capozziello S, Saridakis EN, Bamba K, Sepehri A, Rahaman F, Ali AF, et al. Cosmic space and pauli exclusion principle in a system of m 0-branes. International Journal of Geometric Methods in Modern Physics. 2017;**14**(6):1-32. Article number: 1750095

[23] Antoneli F, Forger M, Gaviria PA, Hornos JEM. On amino acid and codon assignment in algebraic models for the genetic code. International Journal of Modern Physics B. 2010;**24**(4):435-463

[24] Kanjamapornkul K, Pinčák R. Kolmogorov space in time series data. Mathematical Methods in the Applied Sciences. 2016;**39**(15):4463-4483

[25] Capozziello S, Pincak R, Kanjamapornkul K. Anomaly on superspace of time series data. Zeitschrift fur Naturforschung: Section A Journal of Physical Sciences. 2017; **72**(12):1077-1091

[26] Bartoš E, Pinčák R. Identification of market trends with string and D2-brane maps. Physica. 2017;**A479**:57-70

[27] Pinčák R. The string prediction models as invariants of time series in the forex market. Physica. 2013;**A392**: 6414-6426

[28] Pinčák R, Bartoš E. With string model to time series forecasting. Physica. 2015;**A436**:135-146

[29] Kanjamapornkul K, Pinčák R, Bartoš E. The study of Thai stock market across the 2008 financial crisis. Physica. 2016;**A462**:117-133

[30] Capozziello S, Pincak R, Kanjamapornkul K, Saridakis EN. The chern-simons current in systems of dna-rna transcriptions. Annalen der Physik. 2018;**530**(4):1-35. Article number: 1700271

[31] Harley CB, Futcher AB, Greider CW. Telomeres shorten during ageing of human fibroblasts. Nature. 1990;**345**(6274):458-460

[32] Bilson-Thompson S. Braided topology and the emergence of matter. Journal of Physics: Conference Series. 2012;**360**:1-4. Article number: 012056

[33] Pinčák R, Kanjamapornkul K, Bartoš E. Forecasting laurent polynomial in the chern–simons current of v3 loop time series. Annals of Physics. 2019;**531**:1-23. Article number: 1800375

Chapter 8

The Ising Model: Brief Introduction and Its Application

Satya Pal Singh

Abstract

Though the idea to use numerical techniques, in order to solve complex three-dimensional problems, has become quite old, computational techniques have gained immense importance in past few decades because of the advent of new generation fast and efficient computers and development of algorithms as parallel computing. Many mathematical problems have no exact solutions. Depending on the complexity of the equations, one needs to use approximate methods. But there are problems, which are beyond our limits, and need support of computers. Ernst Ising published his PhD dissertation in the form of a scientific report in 1925. He used a string of magnetic moments; spin up (+1/2) and spin down (−1/2), and applied periodic boundary conditions to prove that magnetic phase transition does not exist in one dimensions. Lars Onsager, latter, exactly solved the phase transition problem in two dimensions in 1944. It is going to be a century-old problem now. A variety of potential applications of Ising model are possible now a days; classified as Ising universality class models. It has now become possible to solve phase transition problems in complex three-dimensional geometries. Though the area of spinotronics still needs more engagements of computational techniques, its limited use have provided good insights at molecular scale in recent past. This chapter gives a brief introduction to Ising model and its applications, highlighting the developments in the field of magnetism relevant to the area of solid state physics.

Keywords: surface-directed phase separation, wetting-dewetting, Monte Carlo simulation

1. Introduction

Ernst Ising (**Figure 1**) was born on May 10, 1900, in Loe Koln. He started schooling in 1907 and obtained his diploma at the gymnasium there in the year 1918. After brief military training, he studied mathematics and physics at Gottingen University in the year 1919. After a short gap, he continued his studies and learnt astronomy apart of other subjects. He got focused to theoretical physics at the suggestion of Professor W. Lenz. He started investigating ferromagnetism under supervision of W. Lenz by the end of the year 1922. Ising published short paper in 1925 as a summary of his doctoral thesis [1, 2]. He exactly calculated partition function for one-dimensional lattice system of spins. Ising had first proven that no phase transition to a ferromagnetic ordered state occurs in one dimension at any temperature.

His argument in the favor of his mathematical note was very simple. Suppose, if one of the spins get flipped at a random position because of thermal agitation, there is no force available, which can stop the neighboring spins to flip in the same direction. And this process will go on and on, and completely ordered state will not remain stable at a finite temperature. Thus no phase transition will occur at a finite temperature. Ideally speaking, any ordered state will always remain like a metastable state at finite temperature and nothing more. Molecular motion seizes at absolute zero temperature. So, one may expect that no spin fluctuations may occur at absolute zero temperature. Henceforth, the stable ordered state is a natural outcome at absolute zero temperature. But, it cannot be said to be a critical temperature in true sense. The existence of phase transition at this temperature has no physical meaning, because there is no temperature below it. After going through some approximate calculations, Ising purportedly showed that his model could not exhibit a phase transition in two and three dimensions, either. Latter, his conclusion was proven to be erroneous [1, 2] (**Figure 2**).

Barry Simon has quoted it very well "This model was suggested to Ising by his thesis advisor, Lenz. Ising solved the one-dimensional model, and on the basis of the fact that the one-dimensional model had no phase transition; he asserted that there was no phase transition in any dimension. As we shall see, this is false. It is ironic that on the basis of an elementary calculation and erroneous conclusion, Ising's name has become among the most commonly mentioned in the theoretical physics literature. But history has had its revenge. Ising's name, which is correctly pronounced "E-zing", is almost universally mispronounced "I-zing"."

Ising's paper credited Wilhelm Lenz for his original idea, who had first proposed it in the year 1920. W. Lenz was Ising's research supervisor. It has been often rendered as Lenz-Ising model in many citations. Lenz suggested that dipolar atoms in crystals are free to rotate in quantized manner. He proposed quantum treatment of dipole orientations, though in its classical version, Ising considered only two spin states, i.e., $S = \pm 1/2$. Ising discussed his results with Professor Lenz and Dr. Wolfgang Pauli, who was teaching at Hamburg at that time. Ising's work was first cited by famous contemporary scientist Heisenberg. Heisenberg was first one to realize the failure of Lenz-Ising model. In order to explain ferromagnetism, he developed his own theory, using complicated interactions of spins. There

Figure 1.
Ernst (Ernest) Ising (May 10, 1900–May 11, 1998).

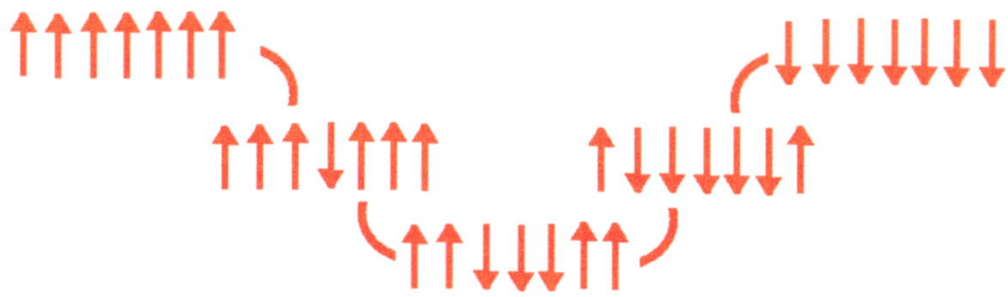

Figure 2.
Random spin flipping in one-dimensional system.

are more scientists in the list, whose contribution to Lenz-Ising model or simply say Ising model must be cited here, because of their historical relevance. They have greatly enriched and contributed to this new model. This list includes scientists like Gorskly (1928), R. H. Fowler (1930), Bragg and Williams (1934), R. Peierls (1936), J. G. Krikwood (1938), Hens Bethe (1939), Kramers and Wannier (1941), and Onsager (1942). They further extended Ising model to a new class of problems.

2. Application of Ising model

Ising model has been extensively used for solving a variety of problems [3–18]. Some of the problems are discussed, here, with appropriate examples.

2.1 Phase separation and wetting/dewetting

Ising model was first exploited for investigating spontaneous magnetization in ferromagnetic film (i.e. magnetization in the absence of external magnetic field). An example case of Ising model using metropolis algorithm is shown in **Figure 3**. Transition temperature depends on the strength of the inter-spin exchange coupling; the dominating term governs the kinetics, when long-range interactions are introduced in the calculations. Latter, it was used to study phase separation in binary alloys and liquid-gas phase transitions (i.e., condensation of molecule in one region of space of the box). Binary alloys constitute of two different atoms. At temperature T = 0, Zn-Cu alloy; known as brass, gets completely ordered. This state is said to be β-brass. In β-brass state, each Zn atom is surrounded by eight copper

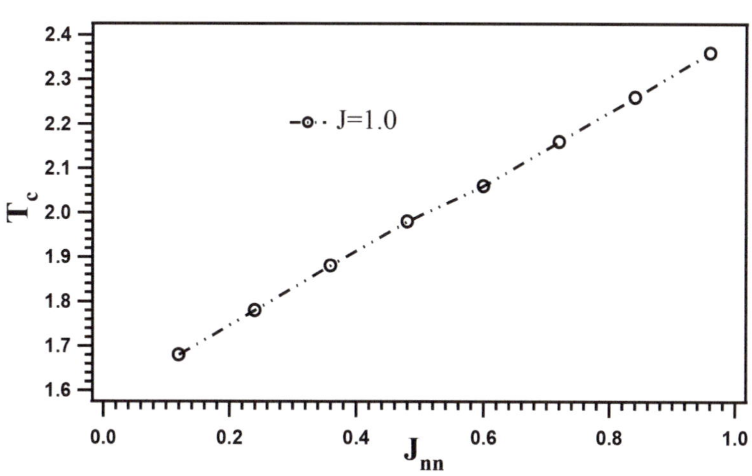

Figure 3.
Variation in critical temperature vs. next nearest exchange coupling for a bcc lattice (reproduced with permission from Singh [3]).

atoms, placed at the corners of the unit cell of the body-centered cubic structure and vice versa. The occupation of each site can be represented by:

$$n_i = \begin{cases} 1 & \text{if site i is occupied by atom A} \\ 0 & \text{if site i is occupied by atom B} \end{cases} \tag{1}$$

The interaction energy between A-A, B-B, and A-B type of atoms are represented by ε_{AA}, ε_{BB}, and ε_{AB}, respectively. Phase separation has been studied vastly, using Ising model [4–6]. A phase is simply a part of a system, separated from the other part by the formation of an interface; that essentially means that two components aggregate and form rich regions of A and B type of molecules with an interface in between them. The evolution of two distinct phases, when an initial random but homogeneous mixture is annealed below a definite temperature, is known as phase separation. Phase separation leads to discontinuity and inhomogeneity in the systems. This happens because the phase-separated regions are energetically more stable. Phase separation has been an old problem and has been extended to study diverse phenomena ranging from magnetic liquid-liquid phase separation to protein-protein phase separation in biological systems. This process has also been studied in the presence of external surfaces having affinity to one type of atom or molecule (**Figure 4a**). Both theoretical and experimental methods have been exploited and have been found in close agreement. Formation of long ridges and circular drops has been reported numerous occasions using lattice-based Ising model. For example, one may look into John W. Cahn research paper published in *The Journal of Chemical Physics* in the year 1965. The TEM image taken for Vycor, in which one phase had been leached out and the voids were filled with lead (**Figure 4b**).

2.2 Lattice-based liquid-gas model

Yang and Lee first coined the term lattice gas in the year 1952. A lattice should have larger volume (V) than the number of lattice molecules (N), so that some of the nodes or lattice vertices are left empty (i.e., N < V). No lattice vertex can be occupied by more than one particle. The interaction potential between two atoms at lattice sites i and j is given by Eq. (2):

$$U(rij) = \begin{cases} -\in & \textit{if } \mathrm{S_i} = \mathrm{S_j} = 1.0 \text{ and rij} = 1.0 \\ \infty & \text{if rij} = 0.0 \\ 0 & \text{else} \end{cases} \tag{2}$$

For surface affinity of lower surface to i^{th} liquid molecule, we chose:

$$V(ri) = \begin{cases} -\dfrac{J_0}{r(i)} & \textit{if } \mathrm{S_i} = 1.0 \\ 0 & \text{otherwise} \end{cases} \tag{3}$$

The occupation number (n_i) of a lattice site i is given by:

$$n_i = \begin{cases} 1 & \text{if site i is occupied} \\ 0 & \text{if site i is un} - \text{occupied} \end{cases} \tag{4}$$

One example case is shown in **Figure 5**. Here, we chose lattice size of $128 \times 128 \times 48$. The fluid-fluid molecule and wall-liquid molecule interactions are defined, respectively, in Eqs. (2) and (3). In canonical ensemble, the three-dimensional lattice is swept one by one; by choosing sites regularly with one of its

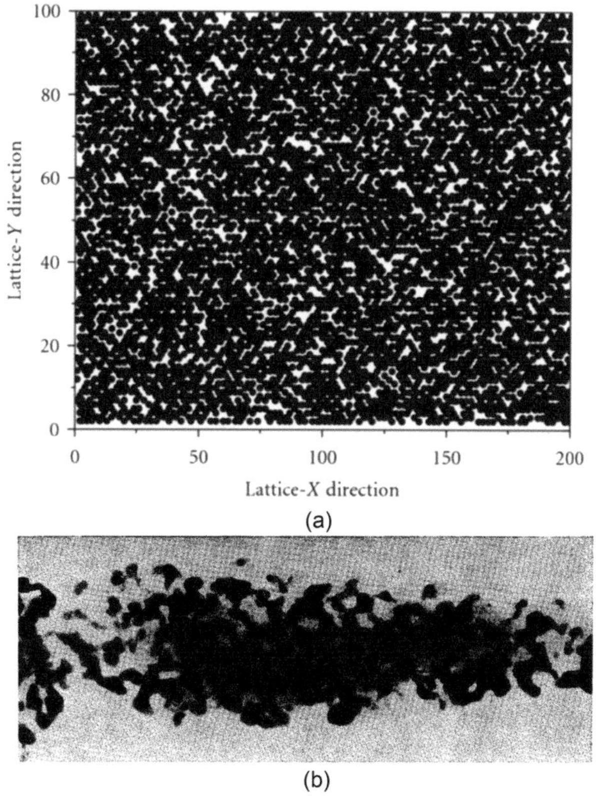

Figure 4.
(a) Surface-directed phase separation and dewetting in conserved binary mixture using two-dimensional lattices of size 200 × 100 nodes. The conserved components are taken in ratio 70:30 at T = 0.70. Majority component is attracted by upper and lower substrates, whereas the minority component has repulsive interaction with the two interfaces. Periodic boundary conditions are applied along X-direction. The micrograph is taken after completion of 30,000 Monte Carlo cycles using Kawasaki exchange method (the figure is reproduced with permission from Singh [5]). (b) Shows Transmission Electron Microscope (TEM) image of unsintered Vycor with one phase replaced by lead (X 200000). Reproduced with permission from W. G. Schmidt and R. J. Charles, Journal of Applied Physics 35, 2552 (1964); doi: 10.1063/1.1702905.

nearest neighbor (i.e., i = n and i + 1 = n + 1). Change in energy is calculated during exchanging of these two sites; the exchange move is accepted, if Exp $[-\Delta E/k_B T]$ is found to be greater than or equal to a random number generated between [0, 1]. For all cases of studies here, $\varepsilon = 1$ and $J_0 = 12.0$, and only the lower surface is functional, while the upper surface has only hard-sphere interaction with the fluid molecules. Average number density for liquid-like molecules is taken as 0.25 [16].

Figure 6 shows micrograph of self-aligned liquid columns. The system evolves from an initial homogeneous mixture of liquid- and gas-like molecules obtained by annealing the system at high temperature for few thousand MC cycles. Dynamic Monte Carlo simulation has been used with continuous but random trial movements of the molecules. The lattice-based Ising model using Eqs. (2) and (3) is also supposed to give same results, at least qualitatively.

2.3 Spin glasses

Crystalline solids possess short- and long-range order along its crystal axes and maintain its periodicity in three dimensions. Liquids possess only short-range order, and its molecules have no long-range correlation. Liquid molecules retain only short-range order. Gases possess neither of the two. These are the three phases, in which any matter may exist. What are the glasses then? Glasses are solids,

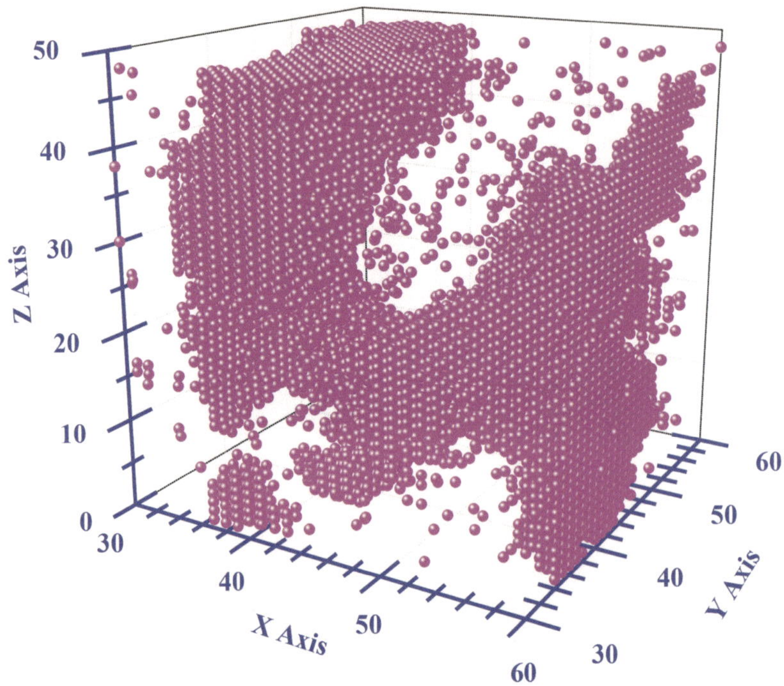

Figure 5.
Micrograph for box thickness H_z = 48 after completion of 20 K M C cycles (figure is reproduced with permission from proceedings, Singh [16]).

possessing no long-range order. Molecules may only locally arrange themselves to minimize its free energy. If the molecular arrangement is completely random, then a term "random media" is assigned to that. Glasses are understood as supercooled liquids. If a liquid is frozen abruptly, so that the molecules do not get sufficient time to organize themselves, some local order can be retained inside the frozen liquid. Glasses have one peculiar property. These retain relatively higher entropy even at quite low temperatures. One example is Mn doped in metals as impurity. Mn atoms interact with other Mn so (i.e. impurity atom) via RKKY interaction $J_{ij}(r) \sim \frac{\cos(2k_F r)}{(k_F r)^3}$. Because of the oscillations in it, the interactions remain random. Such spin systems are classified as spin glasses. There is great deal of frustrations in spin orientations; so on many occasions, these are also referred as "frustrated spin glasses."

Lenz-Ising model did not remain limited to above problems only, but it was extensively used to study liquid mixtures, ternary and quaternary alloys, polymer and their mixtures, random walk problem, and many others. The important aspect of Ising model is that a variety of problems (including some problems mentioned above) can be investigated by the similar kind of modeling and approach all together. It is no longer necessary to develop a different kind of theory for each type of cooperative phenomenon. Despite of all the above, it has been ironical that the inventor of the model, Ernst Ising, gave up the idea on working it, any further presuming that his model has no physical significance. He realized after two decades that he had become famous for his model because of the results obtained by other scientist based on his model, rather by his own work. It has been a queer sensation that the results of Ising model matched with any experimental data or the model was bit artificial. As for as the exponents were concerned, they were of universal nature, and a wide variety of systems have the same Ising exponents. The experimental evidence in favor of it remained a challenge, for many decades. In the

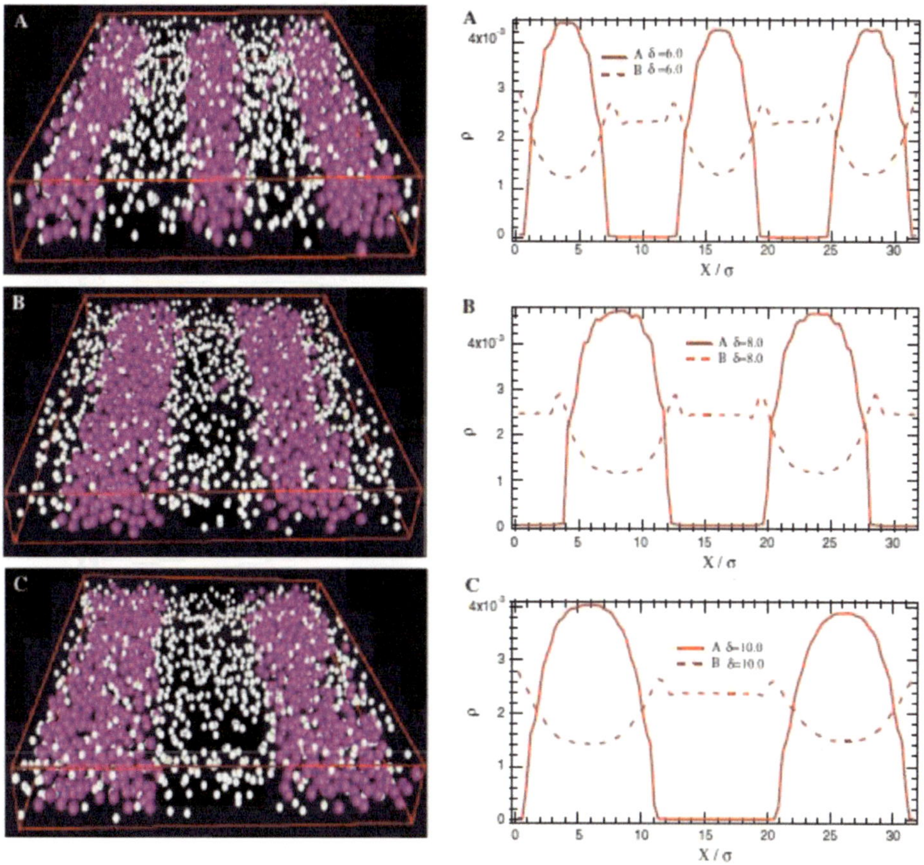

Figure 6.
Self-assembled channels formed in confined geometry; the system starts with a random mixture of square-well fluid (A-type) and hard-sphere (B type) particles. The chemically patterned surface has affinity to (A-type) with interaction range $\lambda_{A\text{-}A}$ = 1.5, $\lambda_{A\text{-}B}$ = 1.5, $\lambda_{Wall\text{-}A}$ = 2.0; *interaction strengths were taken as* ε_{AA} = 1.0, ε_{AB} = 0.5, *and* $\varepsilon_{Wall\text{-}A}$ = 3.0. *Average number density of the system has been taken as* ρ = 0.40. *Pore width H* = 4.0 *and composition ratio A:B* = 50:50 *were taken for all cases of studies. The micrograph and density data were taken after completion of* 40×10^5 *Monte Carlo cycles (the figure is reproduced with permission from Singh et al. [14]).*

year 1974, an alloy was found, which first showed that its magnetic behavior exactly matched with the Onsager result.

3. Mathematical formulation in one dimension

Various textbooks are available nowadays, which discuss Ising model and its applications in greater details [19–22]. Here, brief theory of one-dimensional Ising model is presented. H, Q, and A stands for Hamiltonian, partition function, and free energy of the system, respectively:

$$H_N\{S_i\} = -J\sum_{n,n} S_i S_j - \mu\, B \sum_{i=1}^{N} S_i \tag{5}$$

$$H_N\{S_i\} = -J \sum_{\substack{i=1 \\ <i,j>}}^{N} S_i S_{i+1} - \frac{1}{2}\mu B \sum_{i=1}^{N} (S_i + S_{i+1}) \tag{6}$$

$$Q_N(B,T) = -\sum_{S_1=\pm1} \dots\dots\dots \sum_{S_{N=\pm1}} e^{\beta\sum_i^N\left\{JS_iS_{i+1}+\frac{1}{2}\mu B(S_i+S_{i+1})\right\}} \tag{7}$$

$$\langle S_i|P|S_{i+1}\rangle = e^{\beta\left\{JS_iS_{i+1}+\frac{1}{2}\mu B(S_i+S_{i+1})\right\}} \tag{8}$$

$$Q_N(B,T) = \sum_{S_i=\pm1} \dots\dots\dots \sum_{S_{N=\pm1}} \langle S_1|P|S_2\rangle\langle S_2|P|S_3\rangle \dots\dots\dots \langle S_{N-1}|P|S_N\rangle\langle S_N|P|S_1\rangle \tag{9}$$

$$Q_N(B,T) = \sum_{S_i=\pm1}\sum_{S_{i+1}=\pm1} e^{\beta JS_iS_{i+1}+\frac{1}{2}\mu\beta(S_i+S_{i+1})} \tag{10}$$

$$P = \begin{pmatrix} e^{\beta(J+\mu B)} & e^{-\beta J} \\ e^{-\beta J} & e^{\beta(J-\mu B)} \end{pmatrix} \tag{11}$$

$$Q_N(B,T) = \sum_{S=\pm1}\langle S|P^N|S\rangle = Trace\ (P^N) = \gamma_1^N + \gamma_2^N \tag{12}$$

$$\begin{pmatrix} e^{\beta(J+\mu B)} - \gamma & e^{-\beta J} \\ e^{-\beta J} & e^{\beta(J-\mu B)} - \gamma \end{pmatrix} = 0 \tag{13}$$

$$\gamma^2 - 2\gamma e^{\beta J}\cosh(\beta\mu B) + 2\sinh(2\beta J) = 0 \tag{14}$$

$$\begin{pmatrix} \gamma_1 \\ \gamma_2 \end{pmatrix} = e^{\beta J}\cosh(\beta\mu B) + 2\sinh(2\,\beta J) = 0 \tag{15}$$

$$\begin{pmatrix} \gamma_1 \\ \gamma_2 \end{pmatrix} = e^{\beta J}\cosh(\beta\mu B) \pm \left\{e^{-2\beta J} + e^{2\beta J}\, sinh^2(\beta\mu B)\right\}^{1/2} \tag{16}$$

$$\gamma_2 < \gamma_1;\ \left(\frac{\gamma_2}{\gamma_1}\right)^N \to 0 \tag{17}$$

$$ln\, Q_N(B,T) \cong N ln\, \gamma_1 \tag{18}$$

$$\frac{1}{N}\, ln\, Q_N(B,T) \cong \ln\left[e^{\beta J}\cosh(\beta\mu B) + \left\{e^{-2\beta J} + e^{2\beta J}\, sinh^2(\beta\mu B)\right\}^{1/2}\right] \tag{19}$$

$$A(B,T) = NJ - Nk_BT\ ln\left[\cosh(\beta\mu B) + \left\{e^{-4\beta J} + sinh^2(\beta\mu B)\right\}^{1/2}\right] \tag{20}$$

$$U(B,T) \equiv -k_BT^2\frac{\partial}{\partial T}\left(\frac{A}{k_BT}\right) \tag{21}$$

$$U(B,T) \equiv -NJ - \frac{N\mu B sinh(\beta\mu B)}{\left\{e^{-4\beta J} + sinh^2(\beta\mu B)\right\}^{1/2}} + \frac{2NJe^{-4\beta J}}{\left[\cosh(\beta\mu B) + \left\{e^{-4\beta J} + sinh^2(\beta\mu B)\right\}^{1/2}\right]\left\{e^{-4\beta J} + sinh^2(\beta\mu B)\right\}^{1/2}} \tag{22}$$

Some thermodynamic functions are defined as follows:

$$\overline{M}(B,T) \equiv -\left(\frac{\partial A}{\partial B}\right)_T = \frac{N\mu\ \sinh(\beta\mu B)}{\left[e^{-4\beta J} + sinh^2(\beta\mu B)\right]^{1/2}} \tag{23}$$

$$\chi \equiv \frac{\partial M}{\partial B} \to \chi_0(T) = \frac{N\mu^2}{k_B T} e^{2J/k_B T} \tag{24}$$

3.1 Case A: free boundary with zero field

Partition function is given by:

$$Q\,(0,T) = \sum_{S_1} \dots \dots . \sum_{S_{N-1}} e^{K \sum_{i=1}^{N-1} S_i S_{i+1}} \tag{25}$$

Here, K = β J.
We now define a new variable:

$$\sigma_i = S_i S_{i+1};\, Here\ i = 1, 2, \dots \dots ..N-1 \tag{26}$$

Then we can assign σ to two values, i.e., ±1:

$$\sigma_i = \begin{cases} +1 & if\ \ S_i = S_{i+1} \\ -1 & if\ \ S_i = -S_{i+1} \end{cases} \tag{27}$$

In order to consider contributions from all possible configurations $\{S_1, S_2, S_3 \dots \dots \dots S_N\}$, we need to provide the set of numbers $\{\sigma_1, \sigma_2, \sigma_3 \dots \dots ..\sigma_{N\text{-}1}\}$; here each S_i can take two values as ±1. Configuration in a lattice description means a particular set of values of all spins; if there are N numbers of vertices, there will be 2^N different configurations as a result of permutation and combination of spins. The space, thus formed with these configurations, is called configuration space. Here, summing over σ_i will give only half value of Q, henceforth, we can write:

$$Q\,(0,T) = 2\sum_{\sigma_1} \dots \dots . \sum_{\sigma_{N-1}} e^{K(\sigma_1 + \sigma_2 + \sigma_3 \dots \dots \dots \dots .+\sigma_{N-1})} \tag{28}$$

$$Q\,(0,T) = 2\sum_{\sigma_1} \dots \dots . \sum_{\sigma_{N-1}} e^{K(\sigma_1 + \sigma_2 + \sigma_3 \dots \dots \dots \dots .+\sigma_{N-1})} \tag{29}$$

$$Q\,(0,T) = 2(2coshK)^{N-1} \tag{30}$$

3.2 Case B: periodic boundary with zero field

Now, the partition function is given by:

$$Q\,(0,T) = \sum_{S_1} \dots \dots . \sum_{S_{N-1}} e^{K \sum_{i=1}^{N-1} S_i S_{i+1} + K S_N S_1} \tag{31}$$

Here, $S_{N+1} = S_1$

$$Q\,(0,T) = \sum_{S_1} \dots \dots . \sum_{S_{N-1}} e^{K(\sigma_1 + \sigma_2 + \dots \dots \dots \sigma_{N-1}) + K\sigma_1\sigma_2 \dots \dots \dots \sigma_{N-1}} \tag{32}$$

Since $(S_i)^2 = 1$, we can write $S_1 S_N = S_1.\ S_2.\ S_2.\ S_3.\ S_3 \dots \dots \dots \dots S_{N\text{-}1}.\ S_{N\text{-}1}.\ S_N$

$$Q\,(0,T) = 2\sum_{\sigma_1} \dots \dots . \sum_{\sigma_{N-1}} e^{K(\sigma_1 + \sigma_2 + \dots \dots \dots \sigma_{N-1})} \sum_{n=0}^{\infty} \left(\frac{K\sigma_1\sigma_2\sigma_3 \dots .\sigma_{N-1}}{n!} \right)^n \tag{33}$$

Here, second part in exponential has been converted into a summation series:

$$Q\,(0,T) = 2\sum_{n=0}^{\infty}\frac{K^n}{n!}\left[\sum_{\sigma}\sigma^n e^{K\sigma}\right]^{N-1} \quad (34)$$

$$Q\,(0,T) = 2\sum_{n=0}^{\infty}\frac{K^n}{n!}\left[e^K + (-1)^n e^{-K}\right]^{N-1} \quad (35)$$

$$Q\,(0,T) = (2\cosh \mathrm{K})^N + (2\sinh \mathrm{K})^N \quad (36)$$

It can be shown that in thermodynamic limit, (i.e. $N \to \infty$), the free energy of the system converge to a finite value. Readers are left with the exercise. So, periodic boundary condition, as shown in **Figure 7** (invented by Ising), really helps one to get rid of constructing infinitely large systems. Using appropriate boundary conditions, one may obtain realistic results using large but finite number of spins.

4. Critical phenomena

A lot of research work has been dedicated to observe system behavior near critical points [23–27]. The relevant thermodynamic variables exhibit power-law dependences on the parameter $(T - T_c)$ specifying the distance away from the critical point. The critical points are marked by the fact that different physical quantities pertaining to the system pose singularities at the critical point. These singularities are expressed in terms of power laws of $(T - T_c)$ characterized by critical exponents. As, for example, magnetization $<M>$ identified as an order parameter in magnetism, shows dependence on critical temperature (T_c), with exponent β as follows other exponents are also listed below.

Reduced temperature $t \equiv (T - T_c)/T_c$.
α: specific heat $c\,(t) \sim t^{-\alpha}$; $B \equiv h = 0$.
β: spontaneous magnetization $M\,(t) \sim (-t)^{\beta}, T \le T_c$, $B \equiv h = 0$.
γ: magnetic susceptibility $\chi = \partial M/\partial h, T \sim |t|^{-\gamma}$, $B \equiv h = 0$.
δ: critical Isotherm $M\,(h) \sim |h|^{1/\delta}\ \mathrm{sgn}\,(h), t = 0$.
ν: correlation length, $\xi \sim |t|^{-\nu}$, $B \equiv h = 0$.
η: correlation function $G\,(r) \sim r^{(-d+2-\eta)}, t = 0$, $B \equiv h = 0$.

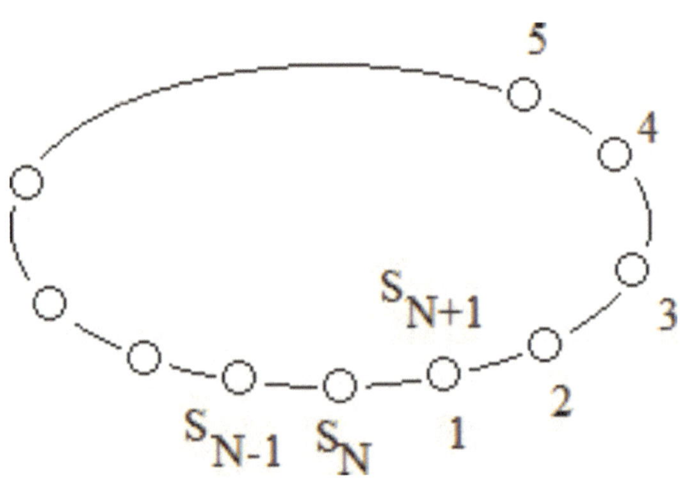

Figure 7.
Representation of periodic boundary conditions in a one-dimensional Ising chain.

4.1 Scaling hypothesis and renormalization group theory

Kadanoff first suggested that, when a system is near critical temperature, individual spins may be grouped into blocks of spins [23]. It is possible because of the fact that the spin-spin correlation length becomes exceedingly large near T_c and details of individual spins no longer remain important. In transformed system, each block plays the role of a single spin. Now, the spin variable associated with a single block is denoted by symbol σ_i. σ_i can take values ± 1. The new system is composed of N' spins (**Figure 8**).

$$N' = l^{-d}N \tag{37}$$

Lattice constant:

$$a' = la \tag{38}$$

In order to preserve the spatial density of the degrees of freedom of spins in the system, the spatial distances are rescaled by the factor l.

$$r' = l^{-1}r \tag{39}$$

Now, the partition function can be updated as follows:

$$Q = \sum_{\{\sigma_i\}} e^{[-\beta H_N\{\sigma_i\}]} \tag{40}$$

This idea was first propounded by Kadanoff, and was later developed by Wilson. This process is also referred as decimation process. A new exchange coupling constant is assigned for interaction between σ_i. This new construction of lattice does not alter the free energy of the system, and it remains the same as obtained by the original method. The rescaling process helps to find relations between various exponents. More detailed discussion on this topic can be found in standard textbooks of *Statistical Mechanics* by Patharia, Huang, etc. Since this process involves

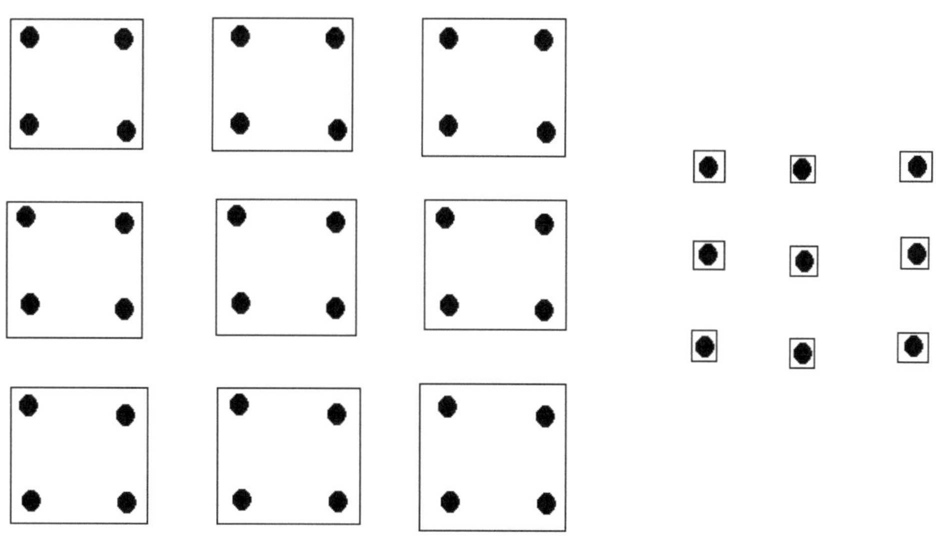

Figure 8.
Spin decimation process in a two-dimensional square lattice. A small cluster of 36 spins gets transformed into 9 nodal points.

length transformation or a change of scale, Wilson introduced the concept of renormalization group theory after removing certain deficiencies in Kadanoff's scaling hypothesis. A greater detail of this is omitted here, because that is beyond the scope of the chapter.

5. Physical realization: simulation results based on Ising model

We now discuss some of the simulation results obtained using Ising model. **Figure 9** shows spontaneous magnetization for a simple cubic crystal (i.e., scc lattice). As the strength of exchange coupling between spin-up and spin-down (J_{AB}) decreases, the critical temperature lowers down. Lower values of J_{AB} weaken the spin flip-flop mechanism; henceforth the system requires further cooling, so that the spin-spin correlation overcomes the fluctuations. Spontaneous magnetization occurs in the absence of external magnetic field [28]. The confirmation of spontaneous process is further confirmed in **Figure 10**. **Figure 10** is plotted for spin

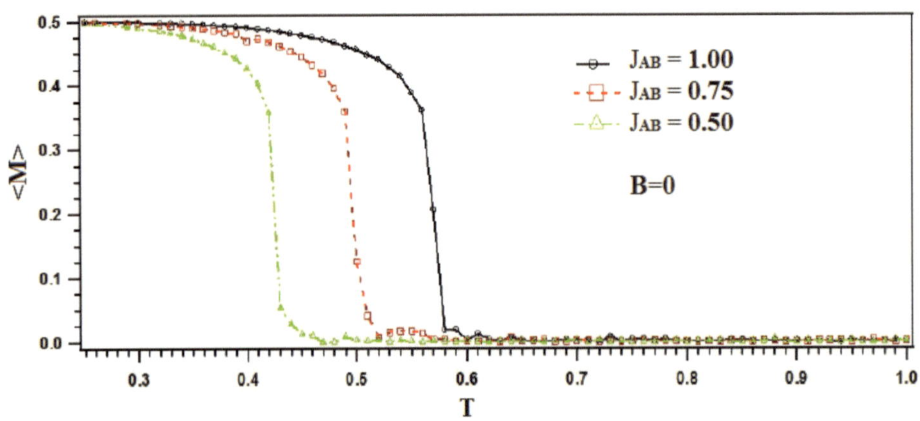

Figure 9.
Spontaneous magnetization in two-dimensional thin film (this figure is reproduced with permission from Singh [28]).

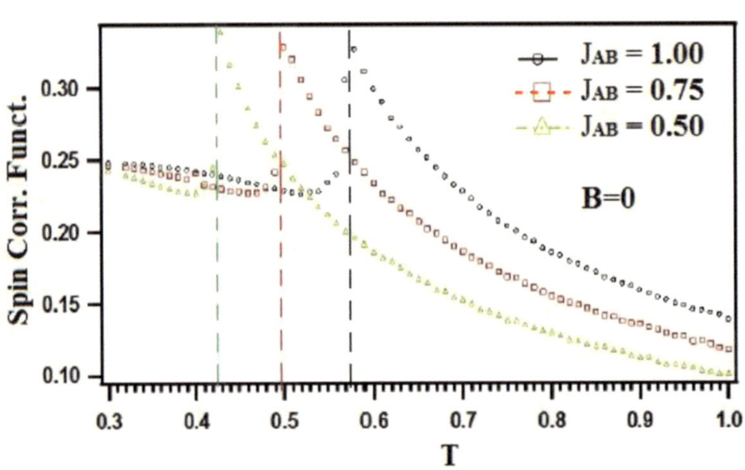

Figure 10.
Correlation function vs. temperature for a two-dimensional thin film. Spontaneous magnetization is marked by discontinuity in it (this figure is reproduced with permission from Singh [28]).

correlation function vs. temperature of the system [28]. The critical temperature is marked by the presence of discontinuity in it. Above critical temperature, the magnetization abruptly falls to zero, which is an indication of paramagnetic state. The critical temperature in ferromagnetic thin film is known as Curie temperature. We observe similar kind of behavior with antiferromagnetic films, though below critical point (also known as Neel temperature), the net average magnetization becomes zero, because opposite spins are energetically favored in this case. The schematic diagram is shown in **Figure 11** [28]. Magnetization vs. external magnetic field curves are plotted in **Figure 12(a)–(d)** for different sets of parameters [28].

Simulation results obtained for a magnetically striped system as schematically shown in **Figure 13** are reported in **Figures 14–17** [29]. One or two alternate rectangular regions are created, using external field. **Figure 14** shows the gradual transition at the interface, where a definite value of external field suddenly gets zero. The spin polarizations in two regions show sharp boundary. The magnetized film, in presence of magnetic field, induces the magnetic zones in proximity where its close external field is zero. Micrograph also indicates for spin-spin phase separation. The corresponding average magnetization vs. temperature and spin correlation function vs. temperature are also plotted in **Figures 15** and **16**, respectively, but these studies are done using Monte Carlo simulation with semi-infinite free boundary conditions. It has been observed that these systems have relatively high critical transition temperatures. **Figure 17** shows the magnetization process with two alternate magnetized zones [29].

Low-dimensional magnetic heterostructures play vital role in spinotronics. Ferromagnets can induce magnetic ordering through a 40-nm-thick amorphous paramagnetic layer, when placed in its close proximity. One has to reconcile with long-range magnetic interaction to correctly measure the extent of induced magnetization. Readers may go through the *Nature Communications* article of F. Magnus et al. published in the year 2016 [17]. The magnetic properties of ferromagnetic materials with reduced dimensions get altered; when the thickness of a film is

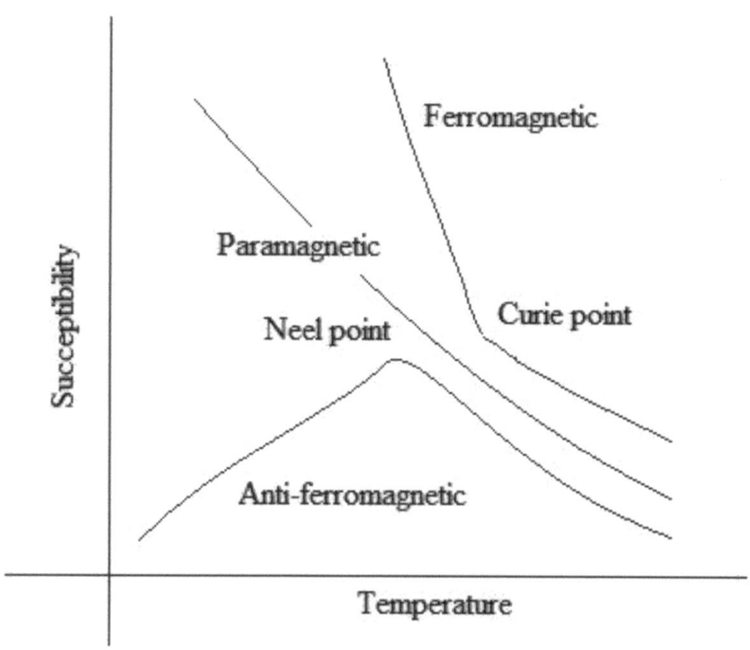

Figure 11.
Schematic representation of ferromagnetic to paramagnetic and antiferromagnetic to paramagnetic transitions.

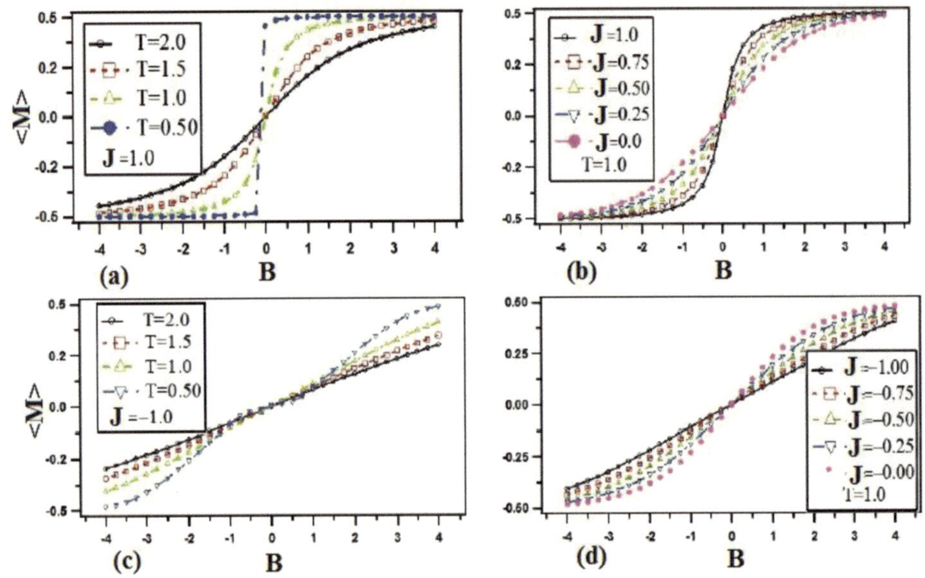

Figure 12.
(a) Magnetization vs. external fields at different temperature T = 0.50, 1.0, 1.5, and 2.0. (b) Magnetization vs. external fields for different exchange couplings J = 0.0, 0.25, 0.50, 0.75, and 1.0. These cases are for ferromagnetic thin films. (c) Magnetization vs. external fields at different temperature T = 0.50, 1.0, 1.5, and 2.0. (d) Magnetization vs. external fields for different exchange couplings J = 0.0, 0.25, 0.50, 0.75, and 1.0. These cases are for antiferromagnetic thin films (this figure is reproduced with permission from Singh [28]).

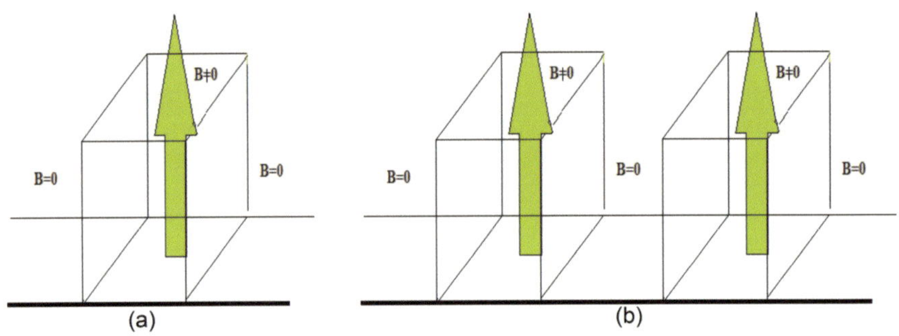

Figure 13.
(a) The system with one slab of size $n_x \times n_y \times n_z = 50 \times 100 \times 100$ exposed to an external magnetic field. (b) The system with two alternate slabs of size $n_x \times n_y \times n_z = 50 \times 100 \times 100$ exposed to an external magnetic field.

reduced below a critical value, the ferromagnetic to paramagnetic transition disappears [18]. Finite-size effects may also weaken or enhance magnetic interactions at the boundaries, as well as restrict the evolution of spin-spin correlation length. Extension of these ideas to model magnetic heterostructures, comprising of multiple magnetic and/or nonmagnetic layers, gives insight into interfacial phenomena. Many current and emerging technologies are based on this central problem. This may be very useful in understanding and exploring problems as metal-insulator transition, which is at the core of many state-of-the-art technologies. Henceforth, computational techniques, especially Ising model, can now be extended to develop and enrich science, for making new technologies. Though, its use can be said at the nascent stage, but with the advancement in computer hardware and efficient algorithms, it's applications in areas related to spinotronics appears to be bright.

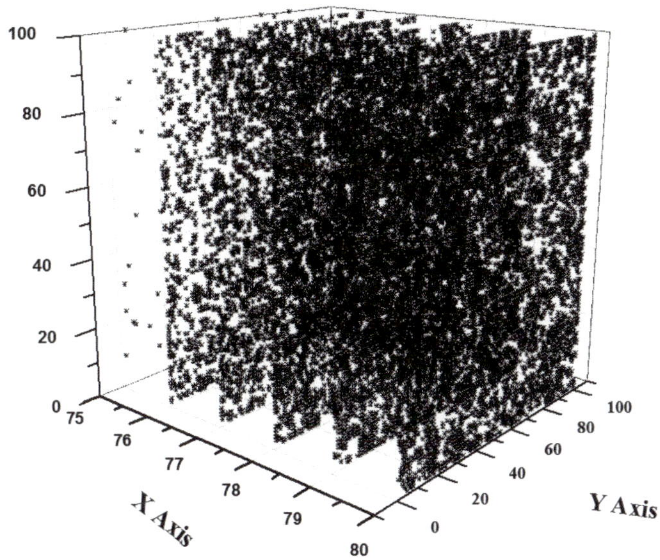

Figure 14.
The micrograph of the coexisting phases in the regions of close proximity of the magnetic barrier indicating for the presence of depletion layer near the barrier.

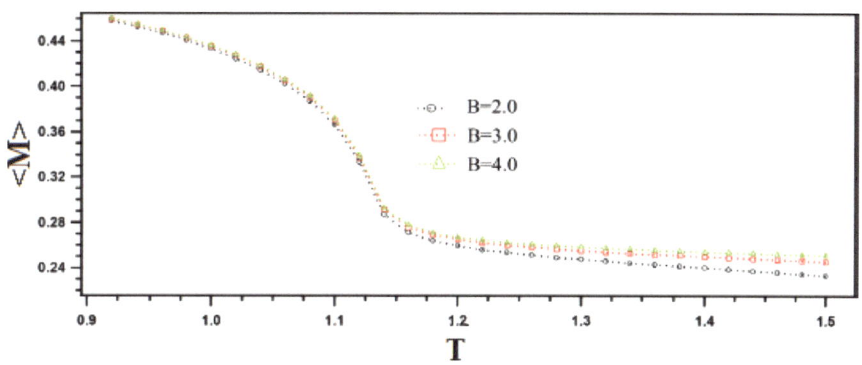

Figure 15.
Magnetization vs. temperature for magnetically striped system. Only one region experiences the presence of external magnetic field as illustrated in ***Figure 12(a)****. This simulation is done for simple cubic lattice with semi-infinite free boundary conditions (the figure is reproduced with permission from Singh [29]).*

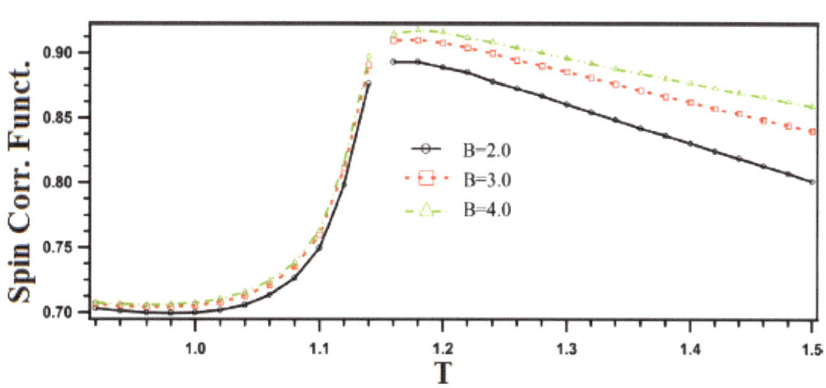

Figure 16.
Spin correlation function vs. temperature for magnetically striped system. Only one region experiences the presence of external magnetic field as illustrated in ***Figure 13(a)****. This simulation is done for simple cubic lattice with semi-infinite free boundary conditions (the figure is reproduced with permission from Singh [29]).*

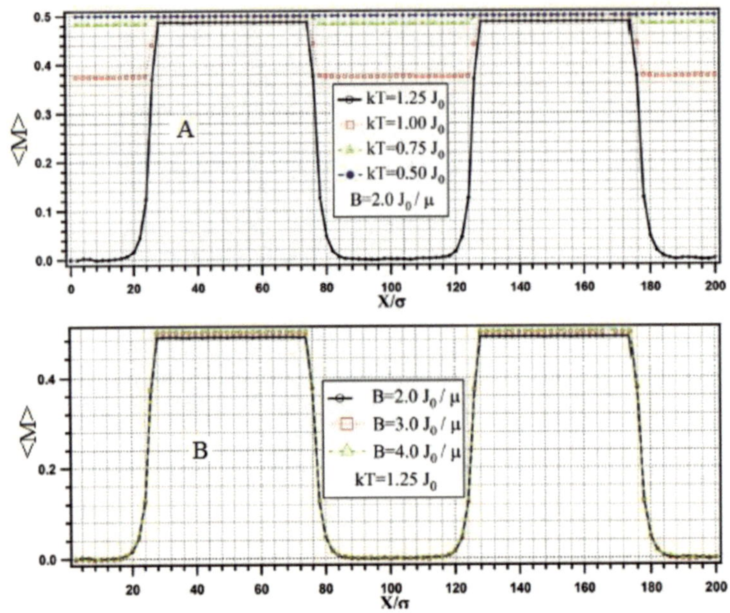

Figure 17.
Magnetization vs. temperature for magnetically striped system. Two alternate regions experience the presence of external magnetic field as illustrated in ***Figure 12(b)****. This simulation is done for simple cubic lattice with semi-infinite free boundary conditions (the figure is reproduced with permission from Singh [29]).*

Author details

Satya Pal Singh
Condensed Matter Physics Research Laboratory, Department of Physics and Material Science, Madan Mohan Malaviya University of Technology, Gorakhpur, Uttar Pradesh, India

*Address all correspondence to: singh.satyapal@hotmail.com

References

[1] Brush SG. History of the Lenz-Ising model. Reviews of Modern Physics. 1967;**39**(4):883-893. DOI: 10.1103/RevModPhys.39.883

[2] Niss M. History of the Lenz–Ising model 1950–1965: From irrelevance to relevance. Archive for History of Exact Sciences. 2009;**63**:243-287. DOI: 10.1007/s00407-008-0039-5

[3] Singh SP. Curie temperature of Ising ferromagnetic film and its dependence on nn exchange coupling. AIP Conference Proceedings. 2018;**1953**: 140015. DOI: 10.1063/1.5033190

[4] Singh SP, Singh K, Roychoudhury M. Monte Carlo simulation for diffusion limited surface directed phase separation. Proceedings of the National Academy of Sciences, India—Section A. 2008;**78**:79-83

[5] Singh SP. Spinodal theory: A common rupturing mechanism in spinodal dewetting and surface directed phase separation (some technological aspects and the significance of dipole-quadrupole interaction in spinodal dewetting). Advances in Condensed Matter Physics. 2011;**2011**:526397. DOI: 10.1155/2011/526397

[6] Singh SP. Spatial correlation in 2D and 3 D thin films of conserved binary mixtures in presence of wetting of substrates by preferred majority component: Interpretation in real scenario. Applied Nanoscience. 2012;**2**: 365-369. DOI: 10.1007/s13204-012-0094-8

[7] Pan AC, Chandler D. Dynamics of nucleation in the Ising model. The Journal of Physical Chemistry. B. 2004; **108**:19681-19686. DOI: 10.1021/jp0471249

[8] Sonsin AF et al. Computational analysis of 3D Ising model using Metropolis algorithms. Journal of Physics: Conference Series. 2015;**630**:012057. DOI: 10.1088/1742-6596/630/1/012057

[9] Katsoulakis MA, Plechac P, Bellet LR. Numerical and statistical methods for the coarse-graining of many-particle stochastic systems. Journal of Scientific Computing. 2008;**37**:43-71. DOI: 10.1007/s10915-008-9216-6

[10] Lundow PH, Markstrom K, Rosengren A. The Ising model for the bcc, fcc and diamond lattices: A comparison. Philosophical Magazine. 2009;**89**(22-24):2009-2042. DOI: 10.1080/14786430802680512

[11] Montroll EW, Potts RB, Ward JC. Correlations and spontaneous magnetization of the two dimensional Ising model. Journal of Mathematical Physics. 1963;**4**:308-322. DOI: 10.1063/1.1703955

[12] Huang R, Gujrati PD. Phase transitions of antiferromagnetic Ising spins on the zigzag surface of an asymmetrical Husimi lattice. Royal Society Open Science. 2019;**6**:181500. DOI: 10.1098/rsos.181500

[13] Singh SP. Revisiting 2D lattice based spin flip-flop Ising model: Magnetic properties of a thin film and its temperature dependence. European Journal of Physics Education. 2014;**5**(3): 8-19

[14] Singh SP, Singh JK, Sharma A. Adsorption of gas like molecules on self-aligned square well fluid channels under confinement of chemically patterned substrates. Applied Nanoscience. 2013;**3**: 179-187. DOI: 10.1007/s13204-012-0118-4

[15] Singh SP. Monte Carlo simulation of microscopic viscosity and rupturing thin polymer film near melt: A molecular

perspective. Physics Letters A. 2017;**381**: 1321-1327. DOI: 10.1016/j.physleta.2017.02.011

[16] SP Singh, A Kumari Singh, Formation of liquid structures and investigation of its interfacial properties using lattice based liquid-gas model, Proceedings of International Conference on Nanoscience and Nanotechnology, 29th Nov.-01st Dec. 2019, VIT, Vellore, Tamil Nadu, India

[17] Magnus F et al. Long-range magnetic interactions and proximity effects in an amorphous exchange-spring magnet. Nature Communications. 2016;**7**:11931. DOI: 10.1038/ncomms119311

[18] Singh SP. Specific heat capacity: Thickness critical spontaneous magnetization in striped ferromagnetic thin films. AIP Conference Proceedings. 2018;**1953**:040020. DOI: 10.1063/1.5032640

[19] Patharia RK, Beale PD. Statistical Mechanics. 3rd ed. Academic Press; 2011

[20] Huang K. Statistical Mechanics, 2nd Ed. Wiley; 2008

[21] Chandler D. Introduction to Modern Statistical Mechanics. USA: Oxford University Press; 1987

[22] Binder K, Heermann D. Monte Carlo Simulation in Statistical Physics, 1868-4513. Berlin Heidelberg: Springer-Verlag;

[23] Kadanoff LP. Scaling Laws for Ising models near Tc*. Physics. 1966;**2**(6): 263-272. DOI: 10.1103/PhysicsPhysiqueFizika.2.263

[24] Holovatch Y. Introduction to renormalization. Condensed Matter Physics. 2006;**9**(2(46)):237-262. DOI: 10.5488/CMP.9.2.237

[25] Kaya T. A new approach to real space renormalization group treatment of Ising model for square and simple cubic lattice. International Journal of Modern Physics B. 2018;**32**(23):1850252. DOI: 10.1142/S0217979218502521

[26] Nauenberg M. Renormalization group solution of the one−dimensional Ising model. Journal of Mathematical Physics. 1975;**16**:703-705

[27] Carneiro CEI, Henriques VB, Salinas SR. Renormalisation group calculations for a spin-1 Ising model with bilinear and biquadratic exchange interactions. Journal of Physics A: Mathematical and General. 1987;**20**: 189-197. DOI: 10.1088/0305-4470/20/1/027

[28] Singh SP. First observations of entropy vs free energy in lattice based modeling for spin coarsening in conserved and non-conserved binary mixtures: The phenomenological study of phase transitions in 2D thin films. Nanosystems: Physics, Chemistry, Mathematics. 2015;**6**(6):882-895. DOI: 10.17586/2220-8054-2015-6-6-882-895

[29] Singh SP. Temperature induced spin polarization in magnetically striped ferromagnetic film: The coexisting phases and the two regime behavior. Macromolecular Symposia. 2017;**376**(1): 1600193. DOI: 10.1002/masy.201600193

Chapter 9

Exchange Bias Effect in Ni-Mn Heusler Alloys

Esakki Muthu Sankaran and Arumugam Sonachalam

Abstract

In this chapter the exchange bias (EB) properties of bulk Mn-rich $Ni_{50-x}Mn_{37+x}Sn_{13}$ $(0 \leq x \leq 4)$ Heusler alloys has been discussed by changing the Ni-Mn concentration. In these alloys the exchange bias field increases with the excess Mn concentration, but exchange bias blocking temperature (T_{EB}) decreases from 149 to 9 K. The hysteresis loop for $Ni_{46}Mn_{41}Sn_{13}$ alloy shows a maximum shift of 377 Oe. The exchange bias property is strongly influenced by varying Ni-Mn concentration in Ni-Mn-Sn alloys then the variation of Mn/Sn. We have observed in these alloys that the T_{EB} would show a decreasing value either by changing the Ni or Sn concentration while the Mn content is above 37% in Ni-Mn-Sn alloys.

Keywords: exchange bias, anisotropy, antiferromagnetism, ferromagnetism, blocking bias temperature, coercive field, exchange bias field

1. Introduction

From the discovery of exchange bias (EB) in CoO particle [1], more work has been done on this area both experimentally and theoretically due to its potential applications in various fields such as spintronic devices, permanent magnets, magnetic recording, read head and giant magnetoresistive sensors, etc. [2–5]. EB arises in the presence of applied magnetic field after cooling the materials and it is connected with the exchange anisotropy formed at the interface between an antiferromagnetic (AFM) and ferromagnetic (FM) materials. The whole phenomenon at low temperature shifts the hysteresis loops along the field axis. This similar kind of phenomenon is observed in multilayer films, small oxide particles, nanostructures and inhomogeneous materials [6–9]. In addition to this, the EB phenomenon is also observed in materials which contains spin glass phase [10]. Recently, Ni-Mn-X (X = Ga, Sb, In, Sn) Heusler-based alloy systems achieved great attention due to their immense applications in magnetic refrigeration, magnetic actuated devices and spintronic devices [11, 12]. The different composition of Ni-Mn-Sn alloy shows a wide physical properties such as magnetic field-induced transition, inverse magnetocaloric effect (IMCE), giant magnetoresistance, giant Hall effect, giant magnetothermal conductivity, magnetic superelasticity effects, exchange bias and shape memory effect [13–17].

The recent observation of EB in the Ni-Mn-based alloys shows an intense interest in the further study of magnetic properties. Due to the different occupations of Mn atoms in the Sn sites as well in the Ni sites, the Ni-Mn-Sn alloy will have excess content of Mn atom. Hence the EB property is very sensitive to the excess Mn. The Ni_2MnSn Heusler alloy crystallizes in $L2_1$ structure, in which the Ni atoms occupy in the (1/2, 1/2, 1/2) and (0, 0, 0) sites, Mn atoms occupy in the (1/4, 1/4, 1/4) site and Sn atoms occupy in the (3/4, 3/4, 3/4) site [18]. In the Mn-rich alloys, the excess Mn occupying Ni and Sn sites couples antiferromagnetically to surrounding Mn atoms on the regular Mn sites [19]. Also the decrease of Mn-Mn distance may lead to AFM exchange between each other in the martensite phase at low temperature. The EB behaviour has been studied in Ni-Mn-X (X = Sb, Sn, In) alloys by several authors [20–23]; particularly in Ni-Mn-Sn alloy, the EB behaviour has been investigated either by varying the Ni/Sn or Mn/Sn concentration [24, 25]. The structural effects, magnetic property and magnetic entropy change have been studied by varying Ni-Mn concentration in the $Ni_{50-x}Mn_{37+x}Sn_{13}$ ($0 \leq x \leq 4$) Heusler alloy system [26]. In the $Ni_{50-x}Mn_{37+x}Sn_{13}$ alloy system, the cubic austenite phase was stabilized by the excess Mn content at room temperature. The martensitic transition temperature decreases from 305 to 100 K by increasing the Mn concentration. The exchange bias blocking temperature (T_{EB}) was found to decrease drastically from 149 to 9 K with increasing Mn concentration. In this work, we have taken up a detailed study on the effect of varying Ni-Mn concentration on EB properties in the bulk $Ni_{50-x}Mn_{37+x}Sn_{13}$ alloys. This chapter explains the EB behaviour by varying Ni-Mn concentration in Ni-Mn-Sn alloys.

2. Experimental details

The compositions of $Ni_{50-x}Mn_{37+x}Sn_{13}$ (x = 0, 1, 2, 3, 4) alloys were prepared by arc melting technique under argon atmosphere. To ensure the homogeneity, the samples are re-melted four times. These alloys were annealed under high vacuum at 1175 K for 6 h and then quenched with Ar gas. The magnetic data were taken suing the physical property measurement system (PPMS-9 T)—vibrating sample magnetometer (VSM) module (Quantum Design, USA). The measurements were taken into two different modes which can be referred as ZFC and FC modes. The sample was firstly cooled in zero magnetic field, and the data was collected by applying a magnetic field of 5 mT during warming in the temperature range of 4–330 K. This refers to zero-field cooled (ZFC) mode. In field cooled (FC) mode, the data was collected without removing the applied field during cooling in the temperature range between 330 and 4 K. Again, the data was recorded upon warming in the range of 4–330 K (referred as field warming (FW)). Magnetization as a function of magnetic field was recorded up to a field of 5 T in the low temperatures.

3. Exchange bias behaviour of $Ni_{50-x}Mn_{37+x}Sn_{13}$ (x = 0, 1, 2, 3, 4) alloys

Figure 1 shows the temperature dependence of magnetization for $Ni_{50-x}Mn_{37+x}Sn_{13}$ (x = 0,1,2,3,4) alloys in an applied magnetic field of 5 mT during zero-field cooling, field cooling and field warming conditions. The curve shows several transitions with thermal hysteresis which has been observed between FC and FW. This thermal hysteresis is the indication of first-order structural transition from austenite to martensite phase. The ferromagnetic transition of austenite phase (T_C^A) occurs at 309 K. The decrease of magnetization below martensite start temperature (M_s) in the FC curve indicates the fractional decrease of austenite phase. The ZFC and FC curves split into two at low temperatures and show a step kind of behaviour in ZFC curve. This specifies that the sample is inhomogeneous magnetically. The transition observed at 120 K is referred as the exchange bias blocking temperature (T_{EB}). The observed magnetic inhomogeneity and T_{EB} in the sample can be attributed to the coexistence of FM and AFM interactions. This kind of antiferromagnetic interaction

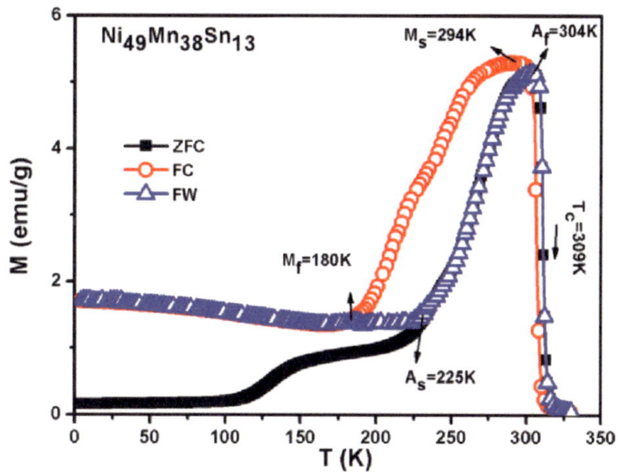

Figure 1.
ZFC, FC and FW thermomagnetic curves for $Ni_{49}Mn_{38}Sn_{13}$ alloy at a field of 5 mT.

occurs from the antiferromagnetic coupling between the Mn atoms in the Ni/Sn sites and Mn atoms in the Mn sites. The similar exchange bias behaviour is observed in further Ni-Mn-X (X = Sb, Sn, In) alloys [20–23]. The presence of AFM interaction in this alloy system was verified by neutron diffraction studies [27].

The ZFC and FC hysteresis loops were measured at 5 K to confirm the exchange bias in $Ni_{50-x}Mn_{37+x}Sn_{13}$ (x = 0, 1, 2, 3, 4) alloys. The ZFC and FC loops were measured from −20 kOe to 20 kOe. To measure the FC loop, the sample was cooled in a field of 50 kOe and recorded the loop from −20 kOe to 20 kOe. **Figure 2** shows the ZFC and FC hysteresis loops for $Ni_{50-x}Mn_{37+x}Sn_{13}$ (x = 0, 1, 2, 3, 4) alloys. The curves from −2 kOe to 2 kOe are shown in the inset in **Figure 3** to see the shift clearly. For all the samples, it is found that the ZFC curve does not exhibit any shift, but the FC curve shifts to the negative field from the origin. This specifies the coexistence of AFM-FM interactions in the sample below room temperatures. The ZFC loop shows a double-shifted loop and is symmetric around zero point, which indicates the existence of FM-AFM coupling. The emergence of double-shifted loop indicates that the different regions of AFM magnetic structure couple to the FM in opposite directions. The Ni-Mn-X (X = Sb, Sn, In) alloys also show similar EB behaviour [20–23].

The temperature dependence of EB was investigated in the temperature range of 5–140 K for $Ni_{50-x}Mn_{37+x}Sn_{13}$ (x = 0, 1, 2, 3, 4) alloys. The temperature interval has been taken as 20 K and the typical FC curves for $Ni_{50-x}Mn_{37+x}Sn_{13}$ (x = 1) alloy are shown in **Figure 3**. It is clear from **Figure 3** that at temperatures of 5 K and 20 K, the hysteresis loops significantly shifted to the negative field which point to the existence of EB in the sample. With the increase of temperature, the hysteresis loop decreases, and finally at 120 K the field shift almost disappears. The temperature where the loop is symmetric can be defined as the T_{EB} for this sample. Moreover, the temperature 120 K nearly coincides with the T_{EB} observed from the thermomagnetic data as shown in **Figure 2**.

Figure 4 shows the values of EB field (H_E) and coercivity (H_C) evaluated from the hysteresis loops at various temperatures for a typical $Ni_{50-x}Mn_{37+x}Sn_{13}$ (x = 1) alloy. The values of EB field and coercivity field are calculated using $H_E = -(H_1 + H_2)/2$ and $H_C = |H_1 - H_2|/2$, respectively, where H_1 and H_2 denote the negative and positive field at which the magnetization equals zero. It is observed from **Figure 4** that with increasing temperature the value of H_E decreases linearly and vanishes around T_{EB}. This validates the EB phenomenon to be real in the

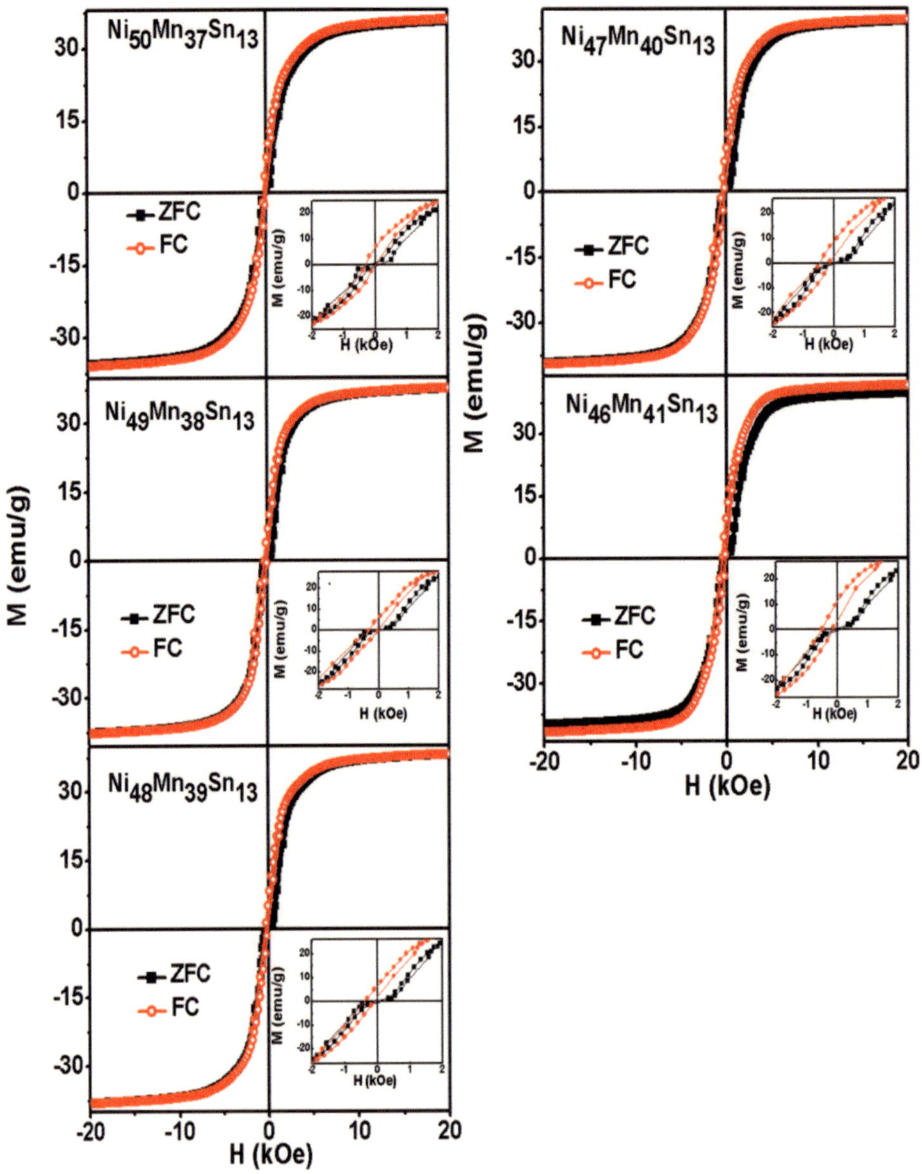

Figure 2.
Magnetic hysteresis loops obtained in the ZFC and FC mode for $Ni_{50-x}Mn_{37+x}Sn_{13}$ (x = 0, 1, 2, 3, 4) alloys at 5 K. inset shows the magnified image in low field range for better visibility of loop shift.

temperatures below T_{EB}. This is because with the increasing temperature, the FM-AFM coupling gets weakened. The disappearance of H_E is due to the domination of FM interaction over the AFM interactions. Conversely, H_C value increases in the beginning with temperature and decreases after reaching a maximum value. Due to the pulling of AFM spins by FM, the coercivity below T_{EB} increases. This arises due to the fact that anisotropy of AFM decreases with increasing temperature.

The T_{EB} values derived from the thermomagnetic curves (not shown for all samples) are plotted with Mn concentration for the alloys in **Figure 5**. From the figure it is seen that with the increase of Mn concentration from 37 to 41% in the $Ni_{50-x}Mn_{37+x}Sn_{13}$ alloy series, the T_{EB} decreases drastically from 149 to 9 K. The large fraction of FM phase at low temperature and the weakening of the

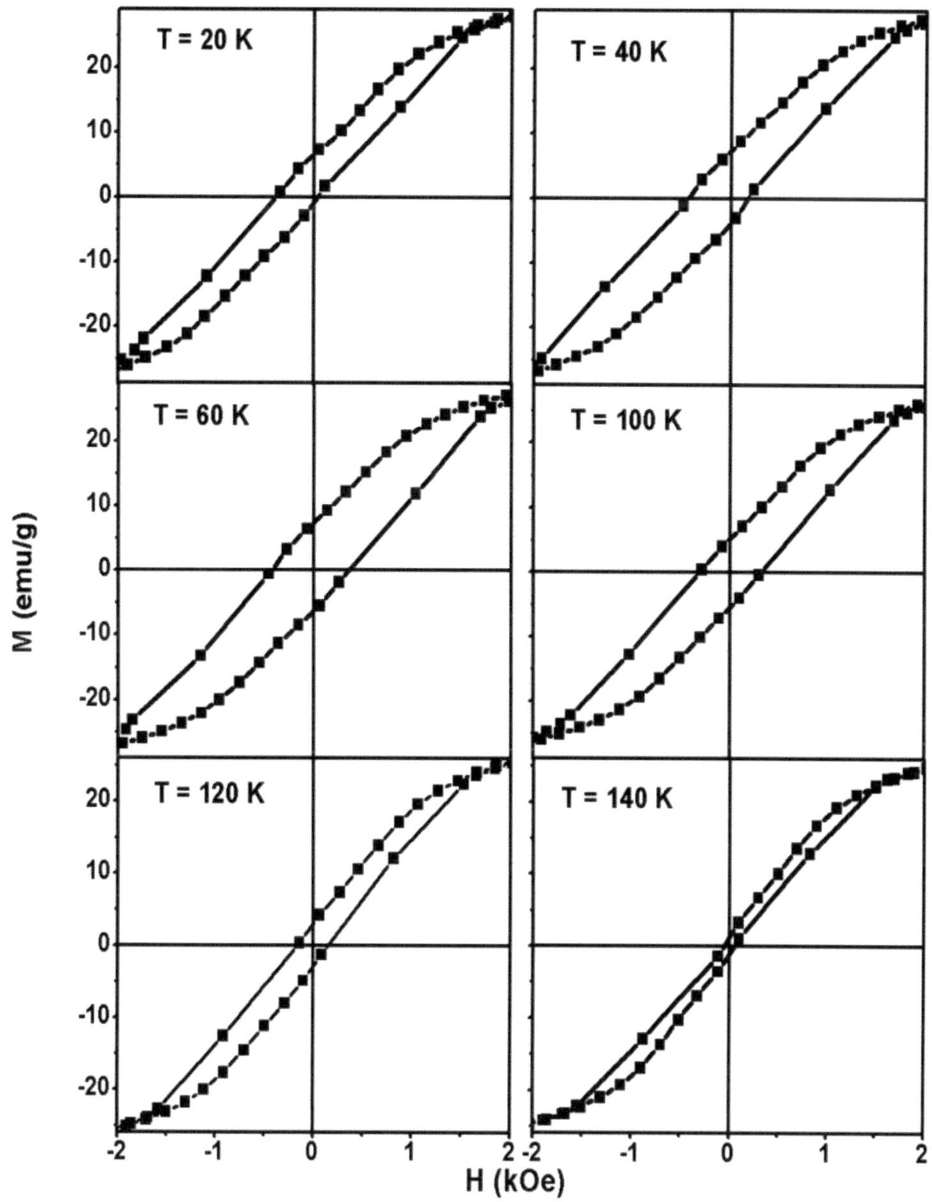

Figure 3.
FC magnetic hysteresis loops measured at various temperatures for $Ni_{49}Mn_{38}Sn_{13}$ alloy. Loop shift decreases with an increase of temperature.

AFM-FM interaction strength were verified by the experimental results which is shown in **Figure 5**, where the increase of Mn concentration increases the saturation magnetization (σ_S) at 5 K marginally. This indicates that the reduction in the FM-AFM interactions occurs due to the increase in Mn content. It can be understood that the weakening of the AFM interactions and the competing AFM and FM interactions led to the decrease of T_{EB} values. Earlier, M. Khan et al. reported the EB properties with varying Mn/Sn concentration in $Ni_{50}Mn_{50-x}Sn_x$ ($11 \leq x \leq 17$) alloys. In his work he reported that the increase of Mn concentration above 37% decreases the T_{EB} value [25]. Same in $Ni_{50-x}Mn_{37+x}Sn_{13}$ alloy series, the T_{EB} value was found to decrease with the increase of Mn content since the Mn

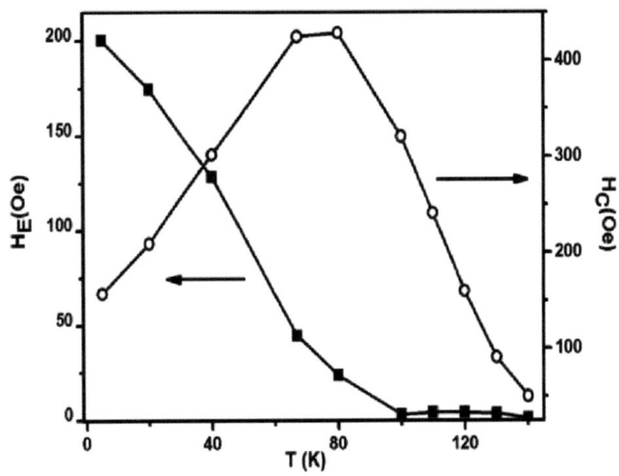

Figure 4.
Variation of EB field (H_E) and coercivity (H_C) with temperature for $Ni_{50-x}Mn_{37+x}Sn_{13}$ (x = 1) alloy.

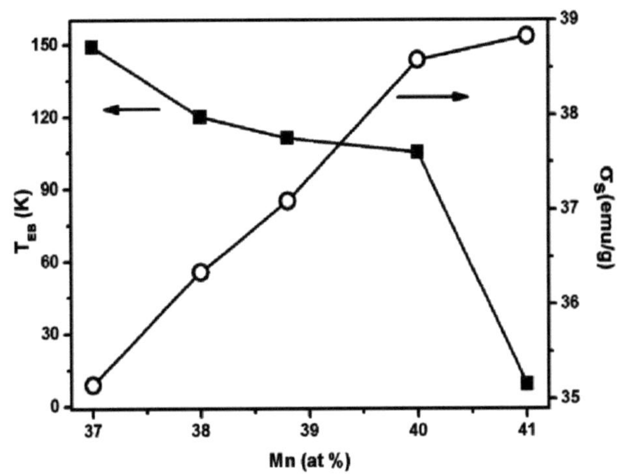

Figure 5.
Variation of T_{EB} and σ_s with Mn concentration in $Ni_{50-x}Mn_{37+x}Sn_{13}$ (x = 0, 1, 2, 3, 4) alloys.

content is more than 37% in $Ni_{50-x}Mn_{37+x}Sn_{13}$ alloy series. These results indicate that the T_{EB} value would decrease either by varying the Sn or Ni content if the Mn content is above 37% in Ni-Mn-Sn alloys.

The variation of H_E with the increase of Mn concentration is shown in **Figure 6** for $Ni_{50-x}Mn_{37+x}Sn_{13}$ alloy series at 5 K (calculated from **Figure 3**). It is noticed that with the increases of Mn concentration from 37 to 41%, the value of H_E increases linearly from 200 to 377 Oe. M. Khan et al. also observed the similar behaviour in Ni-Mn-Sn alloys and found that the H_E increases from ~20 to ~183 Oe with an increase of Mn from 34 to 39% by varying the Mn/Sn concentration [25]. These two results suggest that Ni-Mn variation greatly influences the H_E value in contrast to Mn/Sn variation. Recently, Xuan et al. reported EB by varying Ni/Sn concentration in Ni-Mn-Sn [keeping Mn concentration constant (50%)]. He observed that the increase of Sn content decreases the H_E field [24]. It indicates that the Mn occupying Sn and Ni sites plays a major role in modifying the magnetic interactions and their strength. It is reported that H_E depends on the interface coupling constant and saturation magnetization of the FM [28]. In the present $Ni_{50-x}Mn_{37+x}Sn_{13}$ alloy series, the different magnetic moments and interaction strength arises due to the

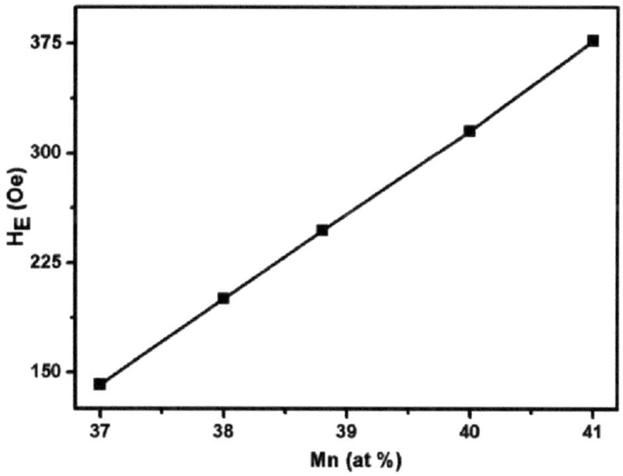

Figure 6.
Variation of EB field (H_E) with Mn concentration for $Ni_{50-x}Mn_{37+x}Sn_{13}$ (x = 0, 1, 2, 3, 4) alloys.

occupation of excess Mn in the Ni sites and Mn atoms occupies Sn and Ni. Hence the increase of H_E with increasing Mn could be due to the increase of interaction coupling strength of different moments in $Ni_{50-x}Mn_{37+x}Sn_{13}$ alloys. In order to understand the magnetic moment of different sites and their magnetic interactions in the sample, detailed neutron diffraction studies are essential.

4. Conclusion

In this chapter, we have discussed the exchange bias behaviour in $Ni_{50-x}Mn_{37+x}Sn_{13}$ (x = 0, 1, 2, 3, 4) alloys by changing the Ni-Mn concentration. The coexistence of AFM and FM exchange interactions is the reason for the EB phenomenon observed in this alloy series. The temperature strongly influences the H_E and H_C. The T_{EB} was found to be same for the Mn content more than 37%, even either varying Ni-Mn or Mn/Sn. With the increase of Mn concentration from 37 to 41%, the value of H_E increases linearly from 200 to 377 Oe in $Ni_{50-x}Mn_{37+x}Sn_{13}$ alloy series.

Acknowledgements

The authors would like to thank Dr. M. Manivelraja and Dr. N.V. Ramarao, DMRL, Hyderabad, India, for providing the facility for material synthesis.

Author details

Esakki Muthu Sankaran[1*] and Arumugam Sonachalam[2]

1 Department of Physics, Faculty of Arts Science and Humanities, Karpagam Academy of Higher Education, Coimbatore, India

2 Centre for High Pressure Research, School of Physics, Bharathidasan University, Tiruchirapalli, India

*Address all correspondence to: esakkimuthu.s@kahedu.edu.in

References

[1] Meiklejohn WH, Bean CP. New magnetic anisotropy. Physics Review. 1956;**102**:1413-1414. DOI: 10.1103/PhysRev.102.1413

[2] Nogués J, Sort J, Langlais V, Skumryev V, Suriñach S, Muñoz JS, et al. Exchange bias in nanostructures. Physics Reports. 2005;**422**:65-117. DOI: 10.1016/j.physrep.2005.08.004

[3] Nowak U, Usadel KD, Keller J, Miltényi P, Beschoten B, Güntherodt G. Domain state model for exchange bias. I. Theory. Physical Review B. 2002;**66**:014430. DOI: 10.1103/PhysRevB.66.014430

[4] Nogués J, Schuller IK. Exchange bias. Journal of Magnetism and Magnetic Materials. 1999;**192**:203-232. DOI: 10.1016/S0304-8853(98)00266-2

[5] Kools JCS. Exchange-biased spin-valves for magnetic storage. IEEE Transactions on Magnetics. 1996;**32**:3165-3184. DOI: 10.1109/20.508381

[6] Stamps RL. Mechanism for exchange bias. Journal of Physics D. 2000;**33**:R247-R268. DOI: 10.1088/0022-3727/33/23/201

[7] Qian T, Li G, Zhang T, Zhou TF, Xiang XQ, Kang XW, et al. Exchange bias tuned by cooling field in phase separated $Y_{0.2}Ca_{0.8}MnO_3$. Applied Physics Letters. 2007;**90**:012503-1-3. DOI: 10.1063/1.2426887

[8] Fraune M, Rüdiger U, Güntherodt G, Cardoso S, Freitas P. Size dependence of the exchange bias field in NiO/Ni nanostructures. Applied Physics Letters. 2000;**77**:3815-3817. DOI: 10.1063/1.1330752

[9] Gruyters M. Spin-glass-like behavior in CoO nanoparticles and the origin of exchange bias in layered CoO/ferromagnet structures. Physical Review Letters. 2005;**95**:077204-1-4. DOI: 10.1103/PhysRevLett.95.077204

[10] Kodama RH, Makhlouf SA, Berkowitz AE. Finite size effects in antiferromagnetic NiO nanoparticles. Physical Review Letters. 1997;**79**:1393-1396. DOI: 10.1103/PhysRevLett.79.1393

[11] Planes A, Manosa L, Acet M. Magnetocaloric effect and its relation to shape-memory properties in ferromagnetic Heusler alloys. Journal of Physics. Condensed Matter. 2009;**21**:233201. DOI: 10.1088/0953-8984/21/23/233201

[12] Cong DY, Roth S, Pötschke M, Hürrich C, Schultz L. Phase diagram and composition optimization for magnetic shape memory effect in Ni–Co–Mn–Sn alloys. Applied Physics Letters. 2010;**97**:021908. DOI: 10.1063/1.3454239

[13] Rama Rao NV, Gopalan R, Chandrasekaran V, Suresh KG. Large low-field inverse magnetocaloric effect near room temperature in $Ni_{50-x}Mn_{37+x}In_{13}$ Heusler alloys. Applied Physics A: Materials Science and Processing. 2010;**99**:265-270. DOI: 10.1007/s00339-009-5517-3

[14] Yu SY, Liu ZH, Liu GD, Chen JL, Cao ZX, Wu GH, et al. Large magnetoresistance in single crystalline $Ni_{50}Mn_{50-x}In_x$ (x=14-16) upon martensitic transformation. Applied Physics Letters. 2006;**89**:162503. DOI: 10.1063/1.2362581

[15] Oikawa K, Ito W, Imano Y, Sutou Y, Kainuma R, Ishida K, et al. Effect of magnetic field on martensitic transition of $Ni_{46}Mn_{41}In_{13}$ Heusler alloy. Applied Physics Letters. 2006;**88**:122507. DOI: 10.1063/1.2187414

[16] Sozinov A, Likhachev AA, Lanska N, Ullakko K. Giant

magnetic-field-induced strain in NiMnGa seven-layered martensitic phase. Applied Physics Letters. 2002;**80**:1746-1748. DOI: 10.1063/1.1458075

[17] Hu FX, Shen BG, Sun JR. Magnetic entropy change in $Ni_{51.5}Mn_{22.7}Ga_{25.8}$ alloy. Applied Physics Letters. 2000;**76**:3460-3462. DOI: 10.1063/1.126677

[18] Kainuma R, Imano Y, Ito W, Sutou Y, Morito H, Okamoto S, et al. Magnetic-field-induced shape recovery by reverse phase transformation. Nature (London). 2006;**439**:957-960. DOI: 10.1038/nature04493

[19] Stager CV, Campbell CCM. Antiferromagnetic order in the Heusler alloy, $Ni_2Mn(Mn_xSn_{1-x})$. Canadian Journal of Physics. 1978;**56**:674-677. DOI: 10.1139/p78-085

[20] Khan M, Dubenko I, Stadler S, Ali N. Exchange bias behavior in Ni–Mn–Sb Heusler alloys. Applied Physics Letters. 2007;**91**:072510-1-3. DOI: 10.1063/1.2772233

[21] Li Z, Jing C, Chen JP, Yuan SJ, Cao SX, Zhang JC. Observation of exchange bias in the martensitic state of $Ni_{50}Mn_{36}Sn_{14}$ Heusler alloy. Applied Physics Letters. 2007;**91**:112505-1-3

[22] Jing C, Chen JP, Li Z, Qiao YF, Kang BJ, Cao SX, et al. Exchange bias behavior and inverse magnetocaloric effect in $Ni_{50}Mn_{35}In_{15}$ Heusler alloy. Journal of Alloys and Compounds. 2009;**475**:1-4. DOI: 10.1016/j.jallcom.2008.07.012

[23] Wang BM, Liu Y, Wang L, Huang SL, Zhao Y, Zhang H. Exchange bias and its training effect in the martensitic state of bulk polycrystalline $Ni_{49.5}Mn_{34.5}In_{16}$. Journal of Applied Physics. 2008;**104**:043916-1-4. DOI: 10.1063/1.2973187

[24] Xuan HC, Cao QQ, Zhang CL, Ma SC, Chen SY, Wang DH, et al. Large exchange bias field in the Ni–Mn–Sn Heusler alloys with high content of Mn. Applied Physics Letters. 2010;**96**:202502-1-3. DOI: 10.1063/1.3428782

[25] Khan M, Dubenko I, Stadler S, Ali N. Exchange bias in bulk Mn rich Ni–Mn–Sn Heusler alloys. Journal of Applied Physics. 2007;**102**:113914-1-4. DOI: 10.1063/1.2818016

[26] Esakki Muthu S, Rama Rao NV, Manivel Raja M, Raj Kumar DM, Mohan Radheep D, Arumugam S. Influence of Ni/Mn concentration on the structural, magnetic and magnetocaloric properties in $Ni_{50-x}Mn_{37+x}Sn_{13}$ Heusler alloys. Journal of Physics D: Applied Physics. 2010;**43**:425002-1-6. DOI: 10.1088/0022-3727/43/42/425002

[27] Aksoy S, Acet M, Deen PP, Manosa L, Planes A. Magnetic correlations in martensitic Ni-Mn-based Heusler shape-memory alloys: Neutron polarization analysis. Physical Review B. 2009;**79**:212401-1-4. DOI: 10.1103/PhysRevB.79.212401

[28] Meiklejohn WH. Exchange anisotropy—A review. Journal of Applied Physics. 1962;**33**:1328-1335. DOI: 10.1063/1.1728716

Chapter 10

Technological Applications of Porphyrins and Related Compounds: Spintronics and Micro-/Nanomotors

David M. Lopes, Juliana C. Araujo-Chaves, Lucivaldo R. Menezes and Iseli L. Nantes-Cardoso

Abstract

The vital role played by porphyrins in cells and their use in therapeutic processes are well known. More recently, the technological applications of porphyrins have attracted the attention of researchers. Porphyrins have the property of half-metallic material, i.e., molecules that can host transition metals making feasible the production of spin-polarized electronic states at different channels. Therefore, porphyrins and hemeproteins are among the materials that have spin-filtering property to be applied in spintronics. Molecular spintronics is an emerging and highly relevant field due to their applications to the development of high-capacity information-storage devices and quantum computers. The catalytic properties of porphyrins and related compounds such as the hemeproteins are also applicable in the fabrication of micro-/nanomotors (MNMs). In this chapter, we describe the advances and future perspectives in the technological applications of porphyrins and related compounds in spintronic devices and micro-/nanomotors.

Keywords: porphyrins, cytochrome *c*, peroxidases, interfaces, advanced materials, micro-/nanomotors, micro-/nanorobots, spintronics, semiconductors, nanotechnology

1. Introduction

1.1 Porphyrins and hemeproteins

Porphyrins are essential compounds for the metabolism of living organisms. Porphyrins result from the substitution of porphine, which is a macrocycle formed by four pyrrole rings linked via methine bridges (**Figure 1a**). The tetrapyrrole ring has space for the coordination of a central transition metal ion with the four nitrogen atoms of the pyrrole rings to form a metalloporphyrin (**Figure 1b**) [1]. The properties of porphyrins can be modulated by substitutions at the β- and *meso*-positions, the central transition metal ions, and the metal ion axial ligands (**Figure 1b**). Another modification of porphyrin ring is the insertion of a carbene in a free-base ring to form the N,N'vinyl-bridged porphyrin and the insertion of a carbene into

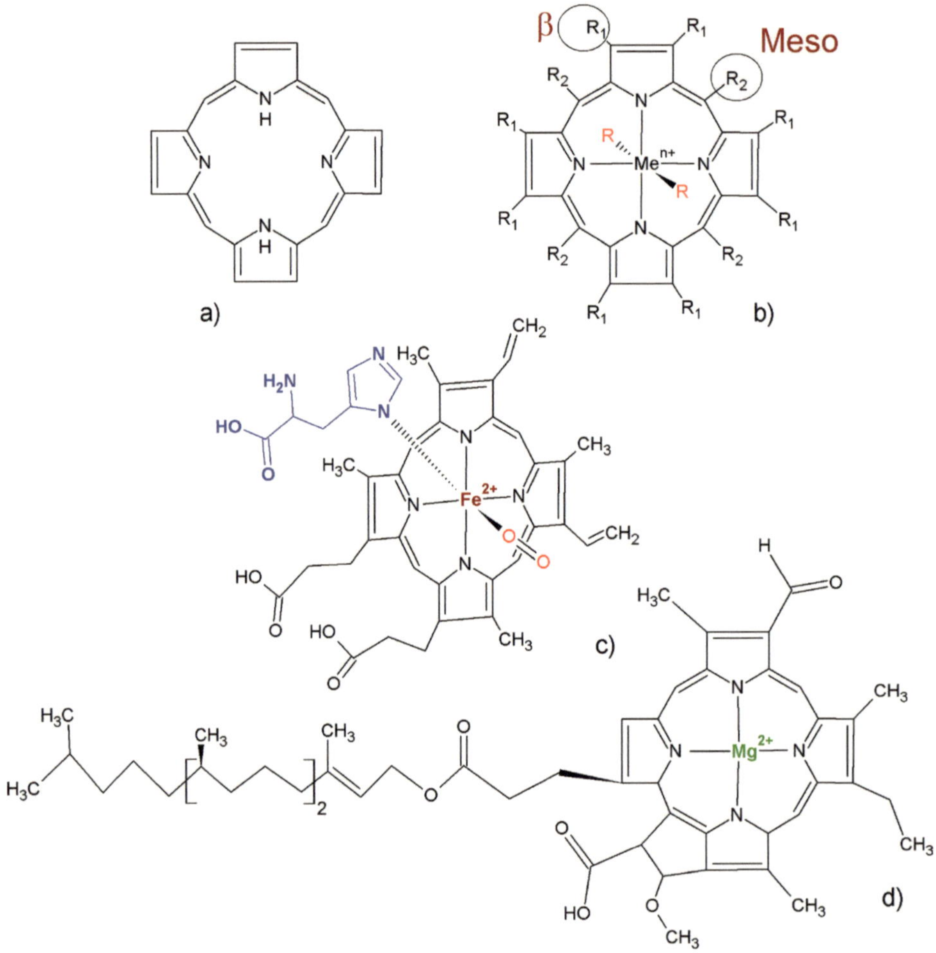

Figure 1.
Porphyrin structure. (a) Porphine; (b) generic structure of a metalloporphyrin with meso- and β-substituents; (c) iron protoporphyrin IX in oxyhemoglobin and oxymyoglobin exhibiting the heme iron axial ligands, lateral chain of histidine at the fifth coordination position, and molecular oxygen at the sixth coordination position; and (d) structure of chlorophyll, a magnesium porphyrin responsible for light harvesting in the photosynthesis process.

the metal-nitrogen bond of a metalloporphyrin [2, 3]. Carbenes can also be added to the porphyrin ring to form homoporphyrin that is also known as expanded porphyrins [4, 5]. The replacement of a nitrogen by C, O, S, Se, and Te results in core-modified porphyrins that are a platform for organometallic chemistry [6]. Two porphyrins are key groups for the energetic metabolism, oxygen transport, and photosynthesis: the iron protoporphyrin IX, the heme group, and chlorophyll (**Figure 1c** and **d**, respectively).

Metalloporphyrins are found in biological systems as the prosthetic group of proteins. Hemeproteins encompass a diversity of proteins associated with the heme group (iron protoporphyrin IX) such as respiratory cytochromes (cyt), cytoglobins (Cgb), neuroglobins (Ngb), myoglobin (Mb), hemoglobin (Hb), cytochrome P450 (CYP), cytochrome b5 (cytb5), and others [7, 8]. The biological activity of hemeproteins is modulated by the microenvironment and iron axial ligands provided by the apoprotein. The modulation of heme iron properties by the microenvironment of proteins results in the same prosthetic group responding for oxygen transport and storage [9], electron transport, $NO^{\bullet}$ trapping, and a variety of catalytic activities such as redox reactions, hydrogen peroxide cleavage, hydroxylation of aromatic

DOI: http://dx.doi.org/10.5772/intechopen.86206

compounds, and others [8]. **Figure 1c** shows the heme group of hemoglobin with histidine imidazole ring as the heme iron axial ligand at the fifth coordination position and molecular oxygen coordinated at the sixth coordination position. Other important biological metalloporphyrins are chlorophylls (magnesium complexes, **Figure 1d**), the plant pigment responsible for plant light harvesting, and cyanocobalamin, a vitamin B12 (cobalt complex, not shown) that participates in the lipid metabolism [10]. The remarkable chemical and photophysical properties of porphyrins have attracted the interest of researchers worldwide [1]. Biological and technological applications of porphyrins can involve the use of native hemeproteins, metallo-substituted hemeproteins, and the product of the tryptic digestion of horse heart cytochrome *c*, microperoxidases [11–19]. Inspired by nature, researchers have synthesized a diversity of nonnatural porphyrins. Theoretical studies of porphyrins have also gained relevance [4, 20–23]. Synthesis of porphyrins is principally motivated by improved use in photodynamic therapy, energy, and catalysis [24–26]. The catalytic and photochemical properties of porphyrins are dependent on the presence and type of the central metal ion with axial ligands, the peripheral decoration, and microenvironment of the ring [11, 27]. In this regard, Zhang et al. [27] demonstrated that the peripheral decoration of porphyrins with simple electron withdrawing and donating groups affects the four Gouterman orbitals with a significant impact on spectroscopic properties and functions (**Figure 2**).

For both solar cells and PDT applications, it is essential that the electron promotion to the lowest excited state can be achieved by the absorption of red light. For energy, the chirality is also interesting because of the chiral-induced spin selectivity (CISS) effect. One example is the generation of hydrogen (H_2) from water splitting by semiconductors. In a standard water splitting system by a semiconductor, the

Figure 2.
Schematic representation of the 18 π electron aromatic ring of a metallated porphyrin with the four nodes of the HOMOs and five nodes of the LUMOs (black-dotted lines). The ML values of HOMO and LUMO pairs are ±4 and ± 5, respectively. The electron density of occupied π MOs is represented by the blue and green shading. The red and yellow shading represents the electron density map of the unoccupied π MOs (molecular orbitals). Scheme inspired in the study of Zhang et al. [27].*

sunlight absorption produces the electron hole pair. The water oxidation by holes (h^+) produces hydroxyl free radicals as intermediates of molecular oxygen evolution. The formation of hydrogen molecules requires that protons (H^+), resulting from the combination of hydroxyl radicals as molecular oxygen, accept the electrons promoted to the conduction band. However, hydrogen gas production competes with the combination of hydroxyl radicals as hydrogen peroxide that is favored by spin-antiparallel photogenerated holes. In the absence of a spin filter, the combination of spin-antiparallel hydroxyl radicals produces singlet molecular oxygen and requires an overpotential of 1 eV, since molecular oxygen is a triplet species in the fundamental state. Chiral molecules act as a spin filter in the electron transfer favoring the production of spin-parallel hydroxyl free radicals and consequently oxygen evolution simultaneously with H_2 production [28]. In the literature, the association of porphyrins and/or hemeproteins with nanostructures, especially for photodynamic therapy purposes, is reported [14, 29]. The reason for this association refers to an enormous quantity of studies and recent findings involving nanostructure properties and manipulation, particularly the potential for drug delivery systems [30]. Nanostructured materials have at least one dimension between 1 and 100 nm. They usually have different (electronic, mechanic, magnetic, optical, etc.) properties from the bulk material, which results in multiple potential applications [31].

1.2 General and basic concepts about nanotechnology, nano-/microrobots (motors), and spintronics

1.2.1 Spintronic

Spintronic concept raised in the late 1980s refers to the use of spins to information transmission and computational operations [32, 33]. Spintronics is an emergent technology grounded in the information transmission by electronic charge and electron spin [34–39]. Spintronic represents a paradigm break in the field of information to combine charge and magnetism in processing and storage. The beginning of spintronics is marked by the discovery of giant magnetoresistance (GMR) effect, in 1988, which resulted in the award of Nobel Prize in Physics in 2007 to Fert and Grunberg [40, 41]. Firstly, spintronic was associated with inorganic oxides, metals, and semiconductors because of the dependence of spin-orbit coupling (SOC). However, organic molecules have wanted properties such as biocompatibility, flexibility, abundance, the possibility of synthesis, low cost [32, 42], and rapidly gained interest in the spintronic studies. The potential applications for spintronics, particularly for electronic devices, are spin filters, spin diodes, spin transistors, spin field-effect transistors, and spin qubits in semiconductor nanostructures [42]. Spintronic has some emerging and promising subfields that are current-induced torque (CIT), spin Hall effect (SHE), spin caloritronics, silicon spintronics, spintronic aspects of graphene and topological insulators (TIs), and chiral-induced spin selectivity effect [32, 34]. The electron spins are degenerate in energy, but the level of degeneracy is broken inside the helix because the electron velocity generates an effective magnetic field that couples with the chiral potential. In a model of DNA double helix, the spin-down electrons aligned preferentially parallel to their velocity in a right-handed helix, while the same occurred with spin-up electrons in the left-handed helix. In an experimental approach, self-assembled monolayers (SAMs) of 3′ thiolated single- and double-strand DNAs (ssDNA and dsDNA, respectively) were attached on a clean 200 nm-thick polycrystalline gold film that was evaporated on glass slides. Photoelectrons were ejected from the gold film by clockwise and counterclockwise circularly polarized light and transmitted through ssDNA and dsDNA monolayers. A more intense transport of electrons ejected with a counterclockwise polarized laser in dsDNA was detected, and no spin selectivity was detected in ssDNA SAMs. Zwang et al. demonstrated that the spin selectivity in DNA is dependent on the supramolecular organization of chiral DNA moieties rather than the chirality of the individual monomers, and thus the spin selectivity can be switched by a conformational change of the molecules [32, 35–39, 43]. The mechanism of CISS effect is believed to be a result of evolution [37], where chiral molecules can increase the conductance of electrons with a spin channel while decreasing the other one [32, 33, 43]. Mishra et al. [44], in recent studies, demonstrated a spin-dependent electron transmission through helical structured bacteriorhodopsin proteins. The study potentially says that the spin degree of freedom may be associated with an important function in electron transport in biological systems. Einati et al. [45] and Roy et al. [46] have shown that the efficiency of electron spin filtering through purple membrane films can be reduced with a green light. So, at potential applications of spin filters, it could modulate the efficiency of the filter.

1.2.2 Nanorobots

Nowadays, a new field of study involving nanotechnology is gaining importance: micro-/nanorobotics. Micro-/nanorobots (MNRs) have autonomous motion provided by micro-/nanomotors (MNMs) that are micro-/nanometer-scale devices powered with the ability to convert chemical, optical, acoustic, magnetic, and electrical energies into mechanical energy [47]. MNRs can be functionalized to perform complex tasks in a microcosm that constitutes the so-called micro-/nanorobots (MNRs) [48]. MNRs have an extensive range of potential applications such as remediation, nanofabrication, repair of materials, engineering, computing, environment monitoring, and especially in theranostics. Drug delivery systems, cell transport, and DNA and RNA insertions are some of the most numerous studies [49, 50]. The size of MNRs allows their application in minimally invasive diagnosis and treatments [51]. There is a basic classification for nanorobots. They can be biological, artificial, or biohybrid [52]. Also, they are classified according to the type of propulsion: self-propelled or external field-propelled ones. The self-propelled nanorobots convert energy from the environment to kinetic energy for independent movement, and it can be done by self-electrophoresis, self-thermophoresis, self-diffusiophoresis, and tiny bubbles [52]. Among the energy sources that self-propelled MNRs can use, light is highly attractive [47]. Light-powered MNMs can obtain energy from an external source and surrounding chemicals to get efficient propulsion through a photocatalytic process and constitute the photocatalytic micro-/nanomotors (PMNMs). Self-propelling PMNMs can be controlled in various ways such as chemical concentration or light intensity [47, 48, 53]. Furthermore, these PMNMs can be operated at low levels of optical and chemical energy input, which are highly desired scenarios. An important aspect is that the photocatalytic reactions of PMNMs can generate the superoxide radicals ($O_2^{-\bullet}$) that give these devices great potential for environmental remediation, especially in the degradation of organic pollutants. The Janus model can be used to explain the basic principles that respond to the self-propulsion of the photocatalytic MNMs (**Figure 3**).

The external field-propelled MNRs depend on an external force such as electric and magnetic field, light impulses, sonic waves, etc. [52]. The fabrication of MNRs can be direct, indirect, or by self-assembly [51]. The techniques used for the MNR fabrication are the same for the regular nanoparticles: top-down (lithography and scanning probe microscopy) and bottom-up (deposition, a solution with reducing agents). The materials used for MNR fabrication could be super magnetic substances, organic and inorganic compounds, and biological substances [51, 54].

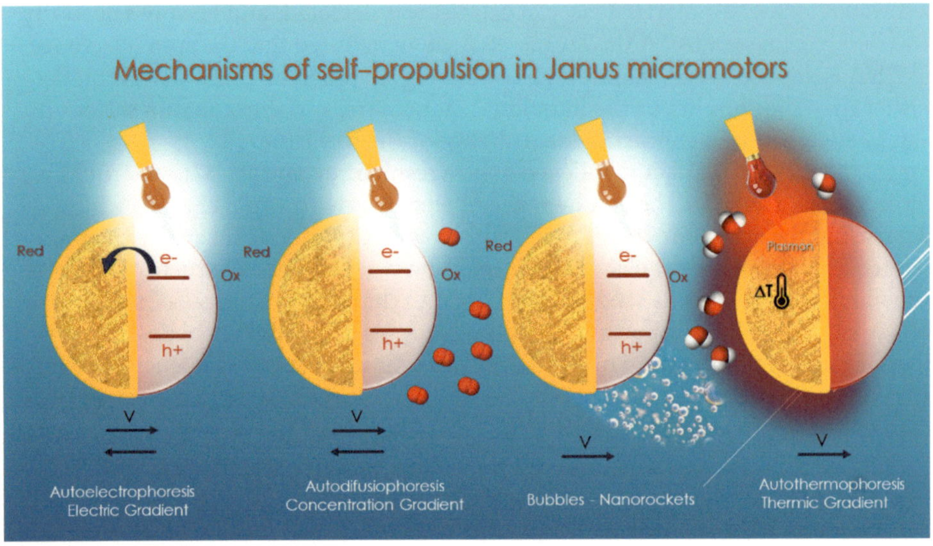

Figure 3.
Self-propelling mechanisms of Janus micromotors. The photocatalytic mechanisms are represented by the three first representations. The formation of a concentration gradient of superoxide ions and bubbles results from the oxidation of hydrogen peroxide in solution. In the right image, irradiation with infrared light on the nanostructured gold layer creates a thermal gradient due to the plasmonic effect. Warming promotes agitation of water molecules that generate one-way movement.

2. Applications of porphyrins and hemeproteins in spintronics

2.1 Porphyrins and derivatives

Single-molecule spintronic devices have gained crescent interest for the use in advanced electronic systems. The question is to find molecular structures which, as single molecule, can exhibit the desirable properties of spintronics such as spin valve [55–60], spin crossover [61–63], spin filtering [64–66], Kondo effect [67], and others. In the literature, porphyrins and derivatives have been described as promising candidates for molecular devices, once they have unique electronic properties [68]. Theoretical and experimental studies on the charge transport of porphyrin-based derivatives have demonstrated desirable physical properties for single-molecule spintronic such as current switching, long-range electron tunneling, current rectifying, and others [55, 69–71]. Several studies have corroborated the potential of porphyrin application in spintronics. Self-assembled porphyrin nanorods showed the mediated conduction through a UHV-STM image with differing HOMO- and LUMO-mediated conductions. The authors demonstrated a conductivity by barrier-type tunneling through distances less than 10 nm and long-distance conduction occurring only through the LUMO band. The self-assembled porphyrin nanorods are an efficient rectifying device that converts alternating current (AC), i.e., a current that periodically inverts direction, to direct current (DC), which moves in a unique direction [72]. In another study, the electronic transport of a nanowire composed by porphyrin-ethyne-benzene conjugates had its effective conductivity assigned to the coplanar conformation of phenyl and porphyrin moieties. The coplanar structure that allows amino or nitro substituent at the meta-position of the phenyl bridge that connects the π-system can provide higher current ratios of the on/off states. The switch effect of meta-substituents in the coplanar conformation disturbs the whole molecule while having only a local impact on the system with a perpendicular conformation. The nanowires formed by π-conjugated systems have potential for switch devices tunable by substituents [69]. Another evidence of porphyrin application in electronic devices was reported by Sedghi et al. [71]. Nanowires formed by porphyrin molecules linearly oligomerized (oligo-porphyrin wires) can mediate temperature-dependent electron transport. The study showed that the system conductance has temperature dependence and it suggests a long-range electron tunneling [71]. For application in spintronic devices, Cho et al. [73] proposed a theoretical organometallic framework formed by one-dimensional infinite chromium porphyrin array in which chromium atoms are located in a straight line (**Figure 4a**). The system exhibited spin filter property when the simulations were carried out with dimeric form, Cr-PA_2, between Au electrodes (**Figure 4b**).

The fabrication of spin-dependent electronics, the spintronic devices, requires the external control of the magnetization of the that behaves like a magnet. In this regard, the paramagnetic porphyrin molecule is a promising active building block for spintronic devices. Wende et al. [74] studied by experimental and theoretical approach paramagnetic iron porphyrin molecules bound on ferromagnetic Ni and Co films on Cu(100). The authors investigated the porphyrin structural orientation and the magnetic coupling with the substrate. The porphyrin molecules associated with the substrate Co or Ni were ordered ferromagnetically. In the device, the magnetic moment of the porphyrin iron could be rotated in the plane and out of the plane by a magnetization reversal of the substrate. In a similar study, Scheybal et al. [75] also associated porphyrins with metallic films and studied X-ray magnetic circular dichroism (XMCD). In this case, the researchers used manganese (III)-tetraphenylporphyrin chloride (MnTPPCl) molecules adsorbed by cobalt substrate

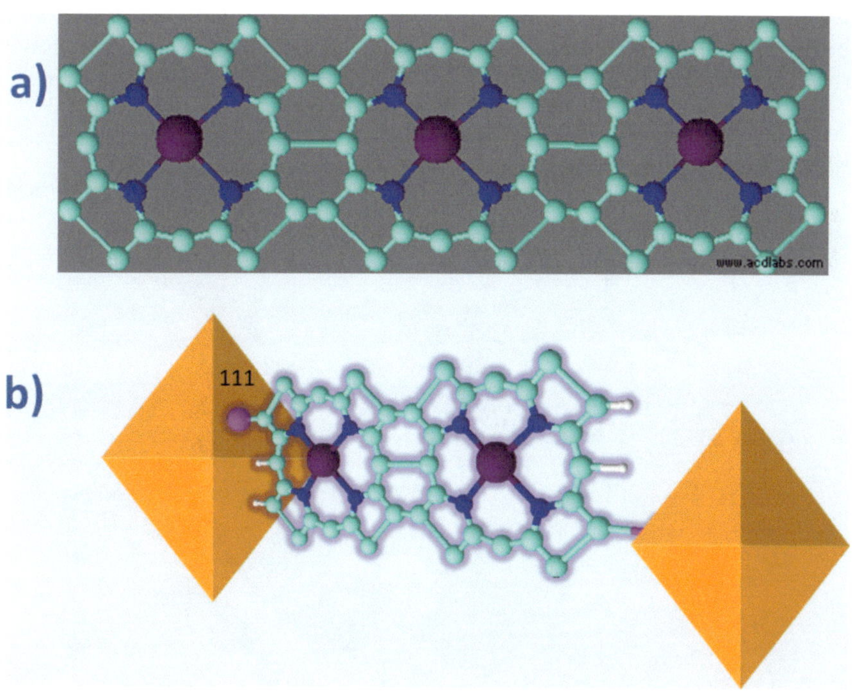

Figure 4.
Examples of nanowires formed by porphyrins. (a) Theoretical organometallic framework built by one-dimensional infinite chromium porphyrin array in which chromium atoms are parallel, the M–Pan; (b) a dimeric form, Cr–PA3, linked to two Au (111) electrodes PA_2 via Au–S bond. Structures of nanodevices modified from Cho et al. [73].

film. The results demonstrated that the film substrate induced a net magnetization on the porphyrin. Chen et al. [76] made calculations of conductance in a ferrous porphyrin. The study showed that the conductance of the iron porphyrin is tuned by mechanical distortion of the porphyrin plane and shifts the coupling state from the low spin to excited spin states. These properties of the system are interesting for sensing applications. Systems containing molecules with switchable spins are promising for the fabrication of materials with spintronic properties. Organometallic molecules such as porphyrins can be switched to on and off magnetic states when associated with the ferromagnetic substrate. Wäckerlin et al. [77] described that cobalt(II)tetraphenylporphyrin (CoTPP) ferromagnetically coupled to nickel thin film (Ni(001)) is switchable from on to off state of Co spin by the complexation with NO that is a spin trans effect. NO coordinates with Co^{2+} leading to the formation of a NO-CoTPP nitrosyl complex that is the off state of the Co spin. The system is restored to the *on* state when NO is thermally dissociated from the nitrosyl complex. Li et al. reported the construction and magnetic characterization of a fully functional system formed by the hybridization of a single magnetic porphyrin molecule with graphene nanoribbons. The fusion of the porphyrin core into graphene through the formation of new carbon rings at chemically predefined positions was demonstrated by scanning tunneling microscopy of high resolution. The authors also demonstrated that porphyrin retains the magnetic functionality and the magnetic anisotropy is modulated by the structure of the contacts [78]. In another study, Lewandowska et al. report a simple and efficient method for the fabrication of porphyrin-graphene oxide hybrids. The hybrid system has donor-acceptor properties and exhibits charge transfer between porphyrin and graphene oxide. The non-covalent interaction between the porphyrin and graphene oxide changes intensely the magnetic properties. The dramatic change in the magnetic properties probably is due to refined tuning of graphene domain magnetism that can be promoted by the modulation electron density produced by electron donor or electron acceptor substituents [79].

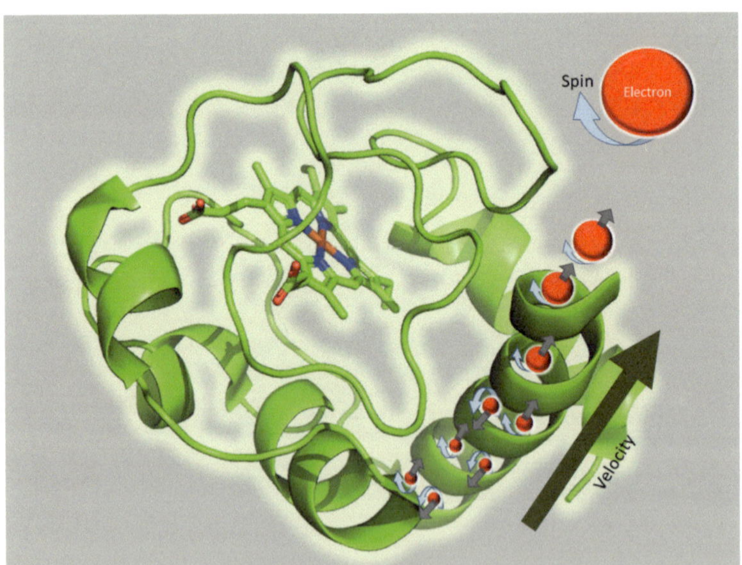

Figure 5.
Chiral-induced spin selectivity (CISS) effect in cytochrome c. cartoon of the spin-filtered electron transport through the chiral a-helix of cytochrome c as reported by Michaeli et al. [82]. Horse heart cytochrome c structure was obtained from protein data Bank, code 1HRC.

2.2 Hemeproteins

The presence of the iron protoporphyrin IX as the prosthetic group of hemeproteins endows these proteins of electronic and magnetic properties that can be applied in spintronics. The hemeproteins have an additional property that is the folding in chiral structures. [80]. The chiral structures such as the α-helices present in cytochrome *c* (**Figure 5**), for instance, can act as spin filters and respond for the chiral-induced spin selectivity (CISS) effect. To date, cytochrome *c* has been the unique hemeprotein used for spin filtering [28, 33, 38].

New types of spin-dependent electrochemistry measurements have been applied to probe the spin-dependent charge transport properties of nonmagnetic chiral molecules such as cytochrome *c*. Besides cytochrome *c*, the photosystems that are complexes of proteins associated with a non-heme porphyrin, the chlorophyll, also have electron transport capacity with spin selectivity [81, 82]. When the measurements were carried out with different orientations of the PSI protein complex, the dependence of spin polarization with the electron transfer path in photosystem I was proven [81, 82].

3. Application of porphyrins and hemeproteins in the construction and working of micro-/nanorobots

3.1 Porphyrins and derivatives

Several studies have been developed to produce nanodevices containing porphyrins with a potential use in MNRs to be applied in theranostics. According to Li et al. [83], porphyrins have a diversity of properties applicable to health preservation, diagnosis, and treatment. Porphyrins can amplify signals for magnetic resonance imaging (MRI), positron emission tomography (PET), infrared fluorescence imaging, and dual modal PET-MRI. Porphyrins have chemical and physical properties that allow the application of these compounds in the detection and destruction of tumors. Porphyrins can efficiently convert light into electronic excitation of molecular oxygen to produce singlet oxygen in photodynamic therapy (PDT) or light to heat for photothermal therapy (PTT). Therefore, porphyrins have been applied in the treatment of solid cancers and ocular vascularization diseases [29]. Also, there are some studies about porphyrin-nanoparticle systems employed in dentistry treatment [84–87]. These systems can be used in the diagnosis of cancer by acting, for instance, as biosensors that exhibit affinity for a single molecule converting biochemical to electrical signals, detection of salivary biomarkers of oral tumors, and others [85, 88]. The capacity of self-assembly in a range of supramolecular aggregates is a crucial property for the application of porphyrins to construct MNRs [29, 89]. Ion et al. [89] demonstrated that porphyrins could self-assemble in several types of supramolecular aggregates such as linear head to tail, J-aggregates, and fractal aggregates with diverse and definite photophysical properties (**Figure 6**) [89].

The study of Ion et al. showed nanotubes formed by porphyrins and the importance of this technique for brain aneurysm instrumentation. They used meso-5, 10, 15, 20-sulfonate-phenyl porphyrin (TPPS4) and observed the formation of organized nanostructures by ionic self-assembly. Neurons and glial cells incubated with porphyrin nanotubes formed interconnected networks featured on the nanotube templates. The capacity of TPPS4 to form nanotubes by self-assembly demonstrates the potential of this porphyrin in the fabrication of NMRs applied to medicine. MNRs must have the capacity to self-propel that could be provided by a diversity of materials and mechanisms. Park et al. describe the fabrication of "swimmers": microstructures with autonomous mobility at water/air interface. The particles of porphyrin-based

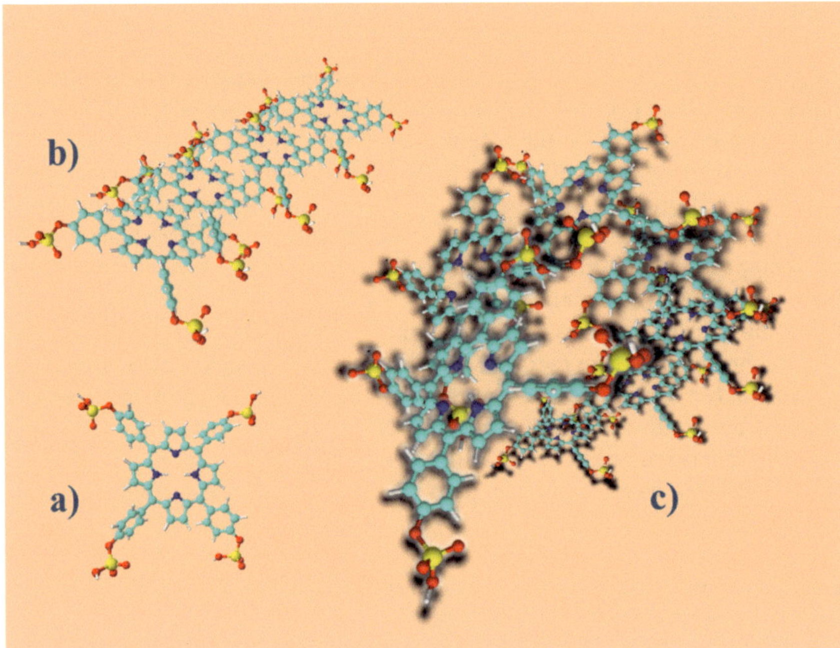

Figure 6.
Supramolecular aggregates of meso-5,10,15,20-sulfonate-phenyl porphyrin (TPPS4). (a) The free-base monomer of TPPS4, (b) the head-to-tail linear self-assembly of TPPS4, and (c) the J-aggregate of TPPS4.

metal–organic frameworks (MOFs) were fabricated with hydrophobic meso-tetra (4-carboxyphenyl)-porphyrin (H4-TCPP-H2, L) ligands bound to Zr-oxo clusters. The H4-TCPP-H2, L responds for the hydrophobic character of the framework [90]. Similar MOFs were described in the literature before, once they are efficient in the controlled release of surface-active substances proportionating a controlled motion. However, usually, the MOFs use high-cost surface-active substances [90]. The MOFs fabricated with meso-tetra(4-carboxyphenyl)-porphyrin ligands bound to Zr-oxo clusters use much less expensive fuels. The particles have the advantage to be refueled multiple times and attained speeds of ca. 200 mm·s^{-1}. Interestingly, the type of fuel, the microstructure, and surface wettability of the MOF surface determine the efficiency of motion. In another study, Serrà et al. [91] reported the fabrication of a multifunctional nanorobotic platform with magnetic properties to promote the death of cancer cells by magnetic and mechanical destruction. A multi-segmented nanowire composed by nickel and gold alternating segments was produced by electrodeposition of metals inside the nanochannels of a polycarbonate membrane. In sequence, the nickel segments were transformed in core-shell Ni/NiO segments by the treatment of the nanowire with NaOH 0.5 M for 6 h. The nanowires were treated sequentially with zinc protoporphyrin IX and 1,9-nonanedithiol that displaces the porphyrin from the gold segments. The nanotubes exhibited ferromagnetism and could be manipulated by a magnet. When the bi-functionalized nanotubes attain cells, magnet or photostimulation can induce cell death that is useful for cancer treatments since the effect of some medical procedures, like hyperthermia and photodynamic therapy, could be improved by application of a rotary magnetic field [91].

3.2 Hemeproteins

Hemeproteins such as hemoglobin (Hb), myoglobin (Mb), horseradish peroxidase (HRP), catalase, and cytochrome *c* (cyt *c*) have the prosthetic group, ferrous or

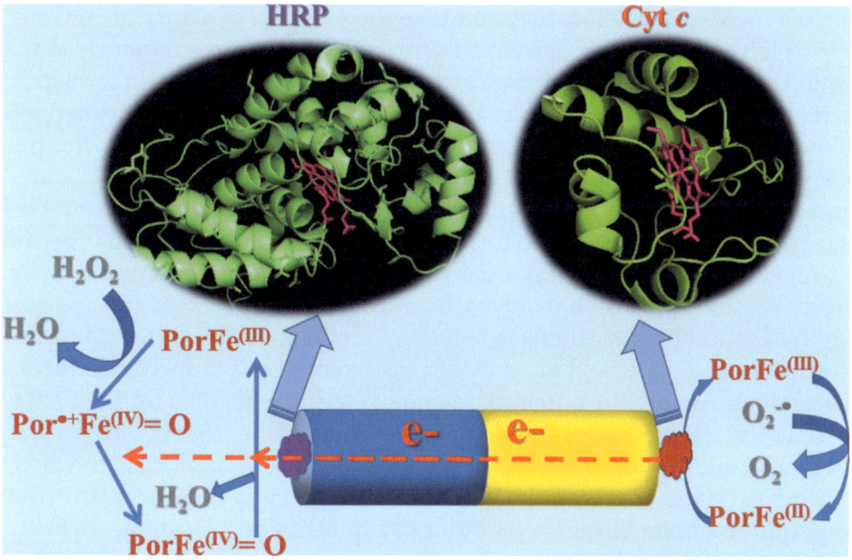

Figure 7.
Proposed bio-electrochemical mechanism behind the enhanced diffusive motion of (HRP)PPy-Au(cyt c) nanorods in $O_2^{-\bullet}$ and H_2O_2 solutions. Native HRP ($PorFe^{(III)}$) reduces H_2O_2 to water and is converted to compound I ($Por^{\bullet+}Fe^{(IV)} = O$), while ferric cyt c ($PorFe^{(III)}$) is reduced by $O_2^{-\bullet}$. Ferrous cyt c ($PorFe^{(II)}$) recycles to the ferric form by transferring electrons through the PPy-Au nanorod to HRP compound I ($Por^{\bullet+}Fe^{(IV)} = O$) and compound II ($PorFe^{(IV)} = O$) that recycles to native HRP. The structures of HRP and cytochrome c were obtained from protein data Bank (1HCH and 1HRC, respectively). Mechanism from Pavel el al. [97].

ferric protoporphyrin IX (heme group), as the redox center. The heme group makes hemeproteins useful for a medical and technological application that involves redox reactions. The use of hemeproteins in nanodevices can be impaired by denaturation or the orientation of the redox site [92]. However, literature has several examples of the use of hemeproteins in nanodevices [93–95]. Hemeproteins can also be used in the self-propelling of MNRs. Hemeproteins can act in MNMs by the bubble recoil mechanism. Catalase is the best hemeprotein for use in MNMs due to the capacity to convert hydrogen peroxide to oxygen generating propulsion bubbles [96]. Pavel et al. [97] fabricated nanorods with self-electrophoresis taking advantage of the combined catalysis of HRP and cytochrome *c* as illustrated in **Figure 7**.

One half of the nanorod was made of polypyrrole (PPy) modified with HRP, and the other half was made of gold and decorated with cyt *c*. [97]. The charge separation was promoted by the reaction of cytochrome *c* with superoxide ion ($O_2^{-\bullet}$) and HRP with H_2O_2. Ferric cyt *c* oxidizes $O_2^{-\bullet}$ to O_2 and is recycled to the oxidized form by transferring one electron through the nanorod to the high valence forms of HRP generated by the reaction with hydrogen peroxide (**Figure 7**) [97]. The study published by Pavel et al. demonstrated that the hemeproteins are robust enough to maintain the activity even immobilized on solid substrates [97].

4. Conclusions and perspectives

Porphyrins and hemeproteins have been widely studied because of their biological roles in energy metabolism and light harvesting in photosynthesis. More recently, with the advancement of bioelectronics and micro-/nanorobotics, porphyrins and hemoproteins have gained interest because of their specific properties. Porphyrins have desirable properties for single-molecule spintronic such as current switching, long-range electron tunneling, current rectifying, and others.

Regarding the hemoproteins, they combine the presence of a porphyrin (iron protoporphyrin IX) as the redox center with the chiral protein structure that acts as a spin filter. To date, cytochrome *c* stands out as the hemoprotein for which the capacity to produce CISS effect has already been demonstrated experimentally. Porphyrins and hemoproteins also have proven potential for nanorobotic application. Porphyrins are particularly useful for nanorobotics applied to medicine because of their photochemical properties. Porphyrins also can self-assemble in structures such as J-aggregates to form nanotubes. On the other hand, the catalytic properties of hemoproteins are the most relevant factor that makes them applicable to self-propulsion in micro-/nanorobotics. The studies and applications of porphyrins and hemoproteins in spintronic and nanorobotic are still in their early stages, and a wide field of study of these compounds is open to the area of bioelectronics. Among the numerous advances that are possible for the field of spintronic, special attention has been given to spinterface, that is, the interface between a ferromagnetic (FM) metal and an organic semiconductor, in which unique hybrid states are formed. The FM metal/molecular interfaces constitute an important building block for the future of spintronics. The unique hybrid states of spinterfaces influence magnetic properties such as magnetic anisotropy, magnetic exchange coupling, interfacial spin polarization, and others. Further, the interactions between the FM metal and organic molecules are tunable in such a way that the spinterfaces are applicable to multifunctional devices meeting the industry tendency of miniaturization using single-molecule devices. The external control of spinterface by external signals, especially light because the ultra-fast optical transmission, is a promising area for future investigations. An important challenge for the design of spintronic devices is the changeable control and switch of single molecules adsorbed on the surface of FM materials. Particularly, for the metalloporphyrins, an interesting example is the use of axial ligands of the porphyrin transition metal center to change the magnetism of the molecular component [98]. $NO^{\bullet}$ was able to reversibly switch the spin state of the Co and Fe of porphyrins adsorbed on Ni(001) Co substrates, respectively. Similarly, NH_3 was able to induce the transition of Ni porphyrin on Co substrate from low to high spin states [99, 100]. Another emerging field of spintronics is the use of antiferromagnets that are affected by spin-polarized currents. Antiferromagnetic materials have several advantages for spintronics such as they do not create external magnetic fields and only weak interactions occur with each other and the antiferromagnets have the characteristic frequencies of switching between their states significantly higher than the values obtained for ferromagnets. Further, the occurrence of ordering in antiferromagnets it is more frequent and occurs at soft conditions than in ferromagnets. Also, these materials can behave as a conductor for a spin polarization and as an insulator for other spin polarization. The antiferromagnets can provide desirable characteristics for spintronics that are high speed of operation in terahertz range, performance, easy manipulation, high sensitivity, and low energy cost [101].

The field of micro/nanorobotics that also can take advantage of the properties of porphyrins and hemeproteins has as the principal challenge for advances as the control and powering of the movement. The crescent interest in the application of MNR in theranostic poses the additional challenge for the use of biocompatible and high-performance materials and fuels. An interesting alternative regarding the elimination of toxic fuels is the use of systems having the propulsion powered by external field that are fuel-free and allow the remote control of the movement. The MNRs with a real potential to operate in vivo are rare now and constitute an important area for future investigations that requires multi- and interdisciplinary studies [102].

DOI: http://dx.doi.org/10.5772/intechopen.86206

Acknowledgements

The author thanks FAPESP 2015/017688-0, 2017/02317-2, SisNano (402289/2013-7), NBB/UFABC, CAPES grant 001, and CNPq (309247/2017-9) for the financial support and CEM/UFABC for the access to facilities.

Conflict of interest

The authors declare no conflict of interest.

Author details

David M. Lopes, Juliana C. Araujo-Chaves, Lucivaldo R. Menezes and Iseli L. Nantes-Cardoso*
Laboratory of Nanostructures for Biology and Advanced Materials, Center of Natural Sciences and Humanities, Federal University of ABC, Santo André, SP, Brazil

*Address all correspondence to: ilnantes@gmail.com

References

[1] Biesaga M, Pyrzyńska K, Trojanowicz M. Porphyrins in analytical chemistry. A review. Talanta. 2000;**51**:209-224

[2] Chan YW, Wood FE, Renner MW, et al. Remarkable disruption of a porphyrin. Insertion of a ruthenium atom into a pyrrole carbon-nitrogen bond of an N,N'-vinyl-bridged porphyrin. Journal of the American Chemical Society. 1984;**106**:3380-3381

[3] Latos-Grazynski L, Cheng RJ, La Mar GN, et al. Reversible migration of an axial carbene ligand into an iron-nitrogen bond of a porphyrin. Implications for high oxidation states of heme enzymes and heme catabolism. Journal of the American Chemical Society. 1981;**103**:4270-4272

[4] Mack J. Expanded, contracted, and isomeric porphyrins: Theoretical aspects. Chemical Reviews. 2017;**117**:3444-3478

[5] Anju KS, Das M, Adinarayana B, et al. *meso* -aryl [20]π homoporphyrin: The simplest expanded porphyrin with the smallest Möbius topology. Angewandte Chemie International Edition. 2017;**56**:15667-15671

[6] Chmielewski PJ, Latos-Grażyński L. Core modified porphyrins—A macrocyclic platform for organometallic chemistry. Coordination Chemistry Reviews. 2005;**249**:2510-2533

[7] Hening WA, Allen RP, Chokroverty S, et al. The iron metabolic system. Restless Legs Syndrome. 2009;**1**:50-60

[8] Lin YW, Wang J. Structure and function of heme proteins in non-native states: A mini-review. Journal of Inorganic Biochemistry. 2013;**129**:162-171

[9] Chakraborti AS. Interaction of porphyrins with heme proteins—A brief review. Molecular and Cellular Biochemistry. 2003;**253**:49-54

[10] Dumon H, Diez M, Nguyen P, et al. Liver lipid metabolism. Journal of Animal Physiology and Animal Nutrition (Berlin). 2008;**92**:272-283

[11] Prieto T, Santana V, AMM B, et al. Structure and catalysis of fe(III) and cu(II) microperoxidase-11 interacting with the positively charged interfaces of lipids. Molecules. 2017;**22**. DOI: 10.3390/molecules22081212 [Epub ahead of print]

[12] Araujo JC, Prieto T, Prado FM, et al. Peroxidase catalytic cycle of MCM-41-entrapped microperoxidase-11 as a mechanism for phenol oxidation. Journal of Nanoscience and Nanotechnology. 2007;7:3643-3652

[13] Prieto T, Marcon RO, Prado FM, et al. Reaction route control by microperoxidase-9/CTAB micelle ratios. Physical Chemistry Chemical Physics. 2006. DOI: 10.1039/b601671e [Epub ahead of print]

[14] Carmona-Ribeiro AM, Prieto T, Nantes IL. Nanostructures for peroxidases. Frontiers in Molecular Biosciences. 2015;**2**. DOI: 10.3389/fmolb.2015.00050 [Epub ahead of print]

[15] Stark BC, Pagilla KR, Dikshit KL. Recent applications of Vitreoscilla hemoglobin technology in bioproduct synthesis and bioremediation. Applied Microbiology and Biotechnology. 2015;**99**:1627-1636

[16] Lebedev N, Trammell SA, Spano A, et al. Conductive Wiring of Immobilized Photosynthetic Reaction Center to Electrode by Cytochrome *c*. 2006. DOI: 10.1021/JA063367Y [Epub ahead of print]

[17] Krainer FW, Glieder A. An updated view on horseradish peroxidases:

Recombinant production and biotechnological applications. Applied Microbiology and Biotechnology. 2015;**99**:1611-1625

[18] McNeil CJ, Smith KA, Bellavite P, et al. Application of the electrochemistry of cytochrome *c* to the measurement of superoxide radical production. Free Radical Research Communications. 1989;7:89-96

[19] Faiella M, Maglio O, Nastri F, et al. De novo design, synthesis and characterisation of MP3, a new catalytic four-helix bundle hemeprotein. Chemistry—A European Journal. 2012;**18**:15960-15971

[20] Rybicka-Jasińska K, Shan W, Zawada K, et al. Porphyrins as photoredox catalysts: Experimental and theoretical studies. Journal of the American Chemical Society. 2016;**138**:15451-15458

[21] Mandal B, Sarkar S, Sarkar P. Theoretical studies on understanding the feasibility of porphyrin-sensitized graphene quantum dot solar cell. Journal of Physical Chemistry C. 2015;**119**:3400-3407

[22] Sirithip K, Prachumrak N, Rattanawan R, et al. Zinc–porphyrin dyes with different meso-aryl substituents for dye-sensitized solar cells: Experimental and theoretical studies. Chemistry—An Asian Journal. 2015;**10**:882-893

[23] Singh A, Lin Y, Quraishi MA, et al. Porphyrins as corrosion inhibitors for N80 steel in 3.5% NaCl solution: Electrochemical, quantum chemical, QSAR and Monte Carlo simulations studies. Molecules. 2015;**2050**:15122-15146

[24] Mugnol KCU, Martins MVA, Nascimento EC, et al. Interaction of Fe(3+)meso-tetrakis (2,6-dichloro-3-sulfonatophenyl) porphyrin with cationic bilayers: Magnetic switching of the porphyrin and magnetic induction at the interface. Theoretical Chemistry Accounts. 2011. DOI: 10.1007/s00214-011-1055-0 [Epub ahead of print]

[25] Araujo-Chaves JC, Kawai C, Melo AFAA, et al. Interaction and reaction of the antioxidant Mn^{III} [Meso-Tetrakis(4-NMethyI Pyridinium)Porphyrin] with the apoptosis reporter lipid Phosphatidylserine. Current Physical Chemistry. 2013;**3**:187-198

[26] Lin VSY, DiMagno SG, Therien MJ. Highly conjugated, acetylenyl bridged porphyrins: New models for light-harvesting antenna systems. Science (80-). 1994;**264**:1105-1111

[27] Zhang A, Kwan L, Stillman MJ. The spectroscopic impact of interactions with the four Gouterman orbitals from peripheral decoration of porphyrins with simple electron withdrawing and donating groups. Organic & Biomolecular Chemistry. 2017;**15**:9081-9094

[28] Zhang W, Banerjee-Ghosh K, Tassinari F, et al. Enhanced electrochemical water splitting with chiral molecule-coated Fe_3O_4 Nanoparticles. ACS Energy Letters. October 2018;**3**(10):2308-2313

[29] Vargas A, Pegaz B, Debefve E, et al. Improved photodynamic activity of porphyrin loaded into nanoparticles: An in vivo evaluation using chick embryos. International Journal of Pharmaceutics. 2004;**286**:131-145

[30] Penon O, Patiño T, Barrios L, et al. A new porphyrin for the preparation of functionalized water-soluble gold nanoparticles with low intrinsic toxicity. ChemistryOpen. 2015;**4**:127-136

[31] Xia Y, Xiong Y, Lim B, et al. Shape-controlled synthesis of metal nanocrystals: Simple chemistry meets complex physics? Angewandte Chemie International Edition. 2009;**48**:60-103

[32] Naaman R, Waldeck DH. Chiral-induced spin selectivity effect. Journal of Physical Chemistry Letters. 2012;**3**:2178-2187

[33] Kumar A, Capua E, Kesharwani MK, et al. Chirality-induced spin polarization places symmetry constraints on biomolecular interactions. Proceedings of the National Academy of Sciences. 2017;**114**:2474 LP-2472478

[34] Sinova J, Žutić I. New moves of the spintronics tango. Nature Materials. 2012;**11**:368-371

[35] Zwang TJ, Hürlimann S, Hill MG, et al. Helix-dependent spin filtering through the DNA duplex. Journal of the American Chemical Society. 2016;**138**:15551-15554

[36] Xie Z, Markus TZ, Cohen SR, et al. Spin specific Electron conduction through DNA oligomers. Nano Letters. 2011;**11**:4652-4655

[37] Gannett PM, Bostick CD, Mukhopadhyay S, et al. Protein Bioelectronics: A Review of What we Do and Do Not Know. Reports on Progress in Physics. 2018;**81**:026601

[38] Mondal PC, Fontanesi C, Waldeck DH, et al. Spin-dependent transport through chiral molecules studied by spin-dependent electrochemistry. Accounts of Chemical Research. 2016;**49**:2560-2568

[39] Gutierrez R, Díaz E, Naaman R, et al. Spin-selective transport through helical molecular systems. Physical Review B. 2012;**85**:81404

[40] Fert A. Nobel lecture: Origin, development, and future of spintronics. Reviews of Modern Physics. 2008;**80**:1517-1530

[41] Baibich MN, Broto JM, Fert A, et al. Giant Magnetoresistance of (001)Fe/(001)Cr magnetic Superlattices. Physical Review Letters. 1988;**61**:2472-2475

[42] Fabian J, Das Sarma S, Žutić I. Spintronics fundamentals and applications. 2004, 2004;**76**. DOI: 10.1103/RevModPhys.76.323 [Epub ahead of print]

[43] Naaman R, Waldeck DH. Spintronics and Chirality: Spin Selectivity in Electron Transport through Chiral Molecules. Annual Review of Physical Chemistry. 2015;**66**:263-284

[44] Mishra D, Markus TZ, Naaman R, et al. Spin-dependent electron transmission through bacteriorhodopsin embedded in purple membrane. Proceedings of the National Academy of Sciences. 2013;**110**:14872-14876

[45] Einati H, Mishra D, Friedman N, et al. Light-controlled spin filtering in bacteriorhodopsin. Nano Letters. 2015;**15**:1052-1056

[46] Roy P, Kantor-Uriel N, Mishra D, et al. Spin-controlled photoluminescence in hybrid nanoparticles purple membrane system. ACS Nano. 2016;**10**:4525-4531

[47] Dong R, Cai Y, Yang Y, et al. Photocatalytic micro/nanomotors: From construction to applications. 2018. DOI: 10.1021/acs.accounts.8b00249 [Epub ahead of print]

[48] Dong R, Wang C, Wang Q, et al. ZnO-based microrockets with light-enhanced propulsion. Nanoscale. 2017;**9**:15027-15032

[49] JGS M, Mayorga-Martinez CC, Wang H, et al. Nano/microrobots meet electrochemistry. Advanced Functional Materials. 2017;**27**. DOI: 10.1002/adfm.201604759 [Epub ahead of print]

[50] Sengupta S, Ibele ME, Sen A. Fantastic voyage: Designing

self-powered nanorobots. Angewandte Chemie International Edition. 2012;**51**:8434-8445

[51] Sharma NN, Mittal RK. Nanorobot movement: Challenges and biologically inspired solutions. International Journal of Smart Sensing and Intelligent Systems. 2008;**1**:87-109

[52] Wang B, Zhang Y, Zhang L. Recent progress on micro- and nano-robots: Towards in vivo tracking and localization. Quantitative Imaging in Medicine and Surgery. 2018;**8**:461-479

[53] Wu Y, Si T, Shao J, et al. Near-infrared light-driven Janus capsule motors: Fabrication, propulsion, and simulation. Nano Research. 2016;**9**:3747-3756

[54] Ummat A, Dubey A, Sharma G, Mavroidis C. Bio Nanorobotics: State of the Art and Future Challenges. In: Yarmush ML, editor. Tissue Engineering and Artificial Organs (The Biomedical Engineering Handbook). CRC Press. 2006:1-42

[55] Yang JF, Zhou L, Han Q, et al. Bias-controlled giant magnetoresistance through cyclopentadienyl-iron multidecker molecules. Journal of Physical Chemistry C. 2012;**116**:19996-20001

[56] Zhu J, Luo Z, Wu S, et al. Magnetic graphene nanocomposites: Electron conduction, giant magnetoresistance and tunable negative permittivity. Journal of Materials Chemistry. 2012;**22**:835-844

[57] Geng R, Luong HM, Daugherty TT, et al. A review on organic spintronic materials and devices: II. Magnetoresistance in organic spin valves and spin organic light emitting diodes. Journal of Science: Advanced Materials and Devices. 2016;**1**:256-272

[58] Geng R, Daugherty TT, Do K, et al. A review on organic spintronic materials and devices: I. magnetic field effect on organic light emitting diodes. Journal of Science: Advanced Materials and Devices. 2016;**1**:128-140

[59] Joshi VK. Spintronics: A contemporary review of emerging electronics devices. Engineering Science and Technology: An International Journal. 2016;**19**:1503-1513

[60] Xu X. A brief review of ferroelectric control of magnetoresistance in organic spin valves. Journal of Materials. 2018;**4**:1-12

[61] Hao H, Zheng X, Song L, et al. Electrostatic spin crossover in a molecular junction of a single-molecule magnet Fe^2. Physical Review Letters. 2012;**108**:017202

[62] Sarkar S, Tarafder K, Oppeneer PM, et al. Spin-crossover in cyanide-based bimetallic coordination polymers—Insight from first-principles calculations. Journal of Materials Chemistry. 2011;**21**:13832

[63] Nakamura M, Takahashi M. Spin crossover in Iron(III) porphyrins involving the intermediate-spin state. In: Mössbauer Spectroscopy. Hoboken, New Jersey: John Wiley & Sons, Inc. pp. 177-201

[64] Huang B, Si C, Lee H, et al. Intrinsic half-metallic BN–C nanotubes. Applied Physics Letters. 2010;**97**:043115

[65] Parida P, Basheer EA, Pati SK. Cyclopentadienyl-benzene based sandwich molecular wires showing efficient spin filtering, negative differential resistance, and pressure induced electronic transitions. Journal of Materials Chemistry. 2012;**22**:14916

[66] Zhao W, Zou D, Sun Z, et al. Mechanical tuning of giant magnetoresistance and spin filtering in manganese diporphyrin-based

molecular junction. ChemElectroChem. 2019;**6**:421-429

[67] Blakesley JC, Neher D. Relationship between energetic disorder and open-circuit voltage in bulk heterojunction organic solar cells. Physical Review B. 2011;**84**:075210

[68] Zeng J, Chen KQ. Spin filtering, magnetic and electronic switching behaviors in manganese porphyrin-based spintronic devices. Journal of Materials Chemistry C. 2013;**1**:4014-4019

[69] An YP, Yang Z, Ratner MA. High-efficiency switching effect in porphyrin-ethyne-benzene conjugates. The Journal of Chemical Physics. 2011;**135**:1-8

[70] Sedghi G, Sawada K, Esdaile LJ, et al. Single molecule conductance of porphyrin wires with ultralow attenuation. Journal of the American Chemical Society. 2008;**130**:8582-8583

[71] Sedghi G, García-Suárez VM, Esdaile LJ, et al. Long-range electron tunnelling in oligo-porphyrin molecular wires. Nature Nanotechnology. 2011;**6**:517-523

[72] Friesen BA, Wiggins B, McHale JL, et al. Differing HOMO and LUMO mediated conduction in a porphyrin nanorod. Journal of the American Chemical Society. 2010;**132**:8554-8556

[73] Cho WJ, Cho Y, Min SK, et al. Chromium porphyrin arrays as spintronic devices. Journal of the American Chemical Society. 2011;**133**:9364-9369

[74] Wende H, Bernien M, Luo J, et al. Substrate-induced magnetic ordering and switching of iron porphyrin molecules. Nature Materials. 2007;**6**:516-520

[75] Scheybal A, Ramsvik T, Bertschinger R, et al. Induced magnetic ordering in a molecular monolayer. Chemical Physics Letters. 2005;**411**:214-220

[76] Chen Y, Prociuk A, Perrine T, et al. Spin-dependent electronic transport through a porphyrin ring ligating an \$\mathrm{Fe}(\mathrm{II})\$ atom: An ab initio study. Physical Review B. 2006;**74**:245320

[77] Wäckerlin C, Chylarecka D, Kleibert A, et al. Controlling spins in adsorbed molecules by a chemical switch. Nature Communications. 2010;**1**:61

[78] Li J, Merino-Díez N, Carbonell-Sanromà E, et al. Survival of spin state in magnetic porphyrins contacted by graphene nanoribbons. Science Advances. 2018;**4**:eaaq0582

[79] Lewandowska K, Rosiak N, Bogucki A, et al. Supramolecular complexes of graphene oxide with porphyrins: An interplay between electronic and magnetic properties. Molecules. 2019;**24**:688

[80] Bentley R. From optical activity in quartz to chiral drugs: Molecular handedness in biology and medicine. Perspectives in Biology and Medicine. 1995;**38**:188-229

[81] Carmeli I, Kumar KS, Heifler O, et al. Spin selectivity in electron transfer in photosystem I. Angewandte Chemie International Edition. 2014;**53**:8953-8958

[82] Michaeli K, Kantor-Uriel N, Naaman R, et al. The electron's spin and molecular chirality—How are they related and how do they affect life processes? Chemical Society Reviews. 2016;**45**:6478-6487

[83] Li Y, Lin T, Luo Y, et al. A smart and versatile theranostic nanomedicine platform based on nanoporphyrin. Nature Communications. 2014;**5**:4712

[84] Babel S, Mathur S. Nanorobotics—Headway towards dentistry. International Journal of Research in Science & Technology. 2011;**1**:3

[85] Divya U. Nanoparticles: An Emerging Trend in Dentistry. 2017 DOI: 10.5005/jp-journals-10037-1117 [Epub ahead of print]

[86] Lyutakov O, Hejna O, Solovyev A, et al. Polymethylmethacrylate doped with porphyrin and silver nanoparticles as light-activated antimicrobial material. RSC Advances. 2014;**4**:50624-50630

[87] González-Delgado JA, Castro PM, Machado A, et al. Hydrogels containing porphyrin-loaded nanoparticles for topical photodynamic applications. International Journal of Pharmaceutics. 2016;**510**:221-231

[88] Konan YN, Gurny R, Allémann E. State of the art in the delivery of photosensitizers for photodynamic therapy. Journal of Photochemistry and Photobiology B: Biology. 2002;**66**:89-106

[89] Ion R-M, Cocina G-C, Iantovics BL, et al. $TPPS_4$ nanotubes architecture for nanorobots with application in cerebral aneurysm. In: AIP Conference Proceedings. 2009;**1117**:201-208

[90] Park JH, Lach S, Polev K, et al. Metal-organic framework 'swimmers' with energy-efficient autonomous motility. ACS Nano. 2017;**11**:10914-10923

[91] Serrà A, Vázquez-mariño G, García-torres J, et al. Magnetic actuation of multifunctional nanorobotic platforms to induce cancer cell death. Death Advanced Biosystems. 2018;**1700220**:1-9

[92] Vidotti M, Carvalhal RF, Mendes RK, et al. Biosensors based on gold nanostructures. 2011;**22**:3-20

[93] Gu H-Y, Yu A-M, Chen H-Y. Direct electron transfer and characterization of hemoglobin immobilized on a Au colloid–cysteamine-modified gold electrode. Journal of Electroanalytical Chemistry. 2001;**516**:119-126

[94] Ju H, Liu S, Ge B, et al. Electrochemistry of cytochrome *c* immobilized on colloidal gold modified carbon paste electrodes and its electrocatalytic activity. Electroanalysis. 2002;**14**:141-147

[95] Zhao X, Mai Z, Kang X, et al. Clay–chitosan–gold nanoparticle nanohybrid: Preparation and application for assembly and direct electrochemistry of myoglobin. Electrochimica Acta. 2008;**53**:4732-4739

[96] Bunea A-I, Pavel I-A, David S, et al. Modification with hemeproteins increases the diffusive movement of nanorods in dilute hydrogen peroxide solutions. Chemical Communications. 2013;**49**:8803-8805

[97] Pavel IA, Bunea AI, David S, et al. Nanorods with biocatalytically induced self-electrophoresis. ChemCatChem. 2014;**6**:866-872

[98] Sun M, Mi W. Progress in organic molecular/ferromagnet spinterfaces: Towards molecular spintronics. Journal of Materials Chemistry C. 2018;**6**:6619-6636

[99] Ormaza M, Abufager P, Verlhac B, et al. Controlled spin switching in a metallocene molecular junction. Nature Communications. 2017;**8**:1974

[100] Wäckerlin C, Tarafder K, Girovsky J, et al. Ammonia coordination introducing a magnetic moment in an on-surface low-spin porphyrin. Angewandte Chemie International Edition. 2013;**52**:4568-4571

[101] Jungfleisch MB, Zhang W, Hoffmann A. Perspectives of antiferromagnetic spintronics. Physics Letters A. 2018;**382**:865-871

[102] Soto F, Chrostowski R. Frontiers of medical micro/nanorobotics: In vivo applications and commercialization perspectives toward clinical uses. Frontiers in Bioengineering and Biotechnology. 2018;**6**:170

Section 4

Mechanics of Solid Deformable Bodies

Chapter 11

2D Elastostatic Problems in Parabolic Coordinates

Natela Zirakashvili

Abstract

In the present chapter, the boundary value problems are considered in a parabolic coordinate system. In terms of parabolic coordinates, the equilibrium equation system and Hooke's law are written, and analytical (exact) solutions of 2D problems of elasticity are constructed in the homogeneous isotropic body bounded by coordinate lines of the parabolic coordinate system. Analytical solutions are obtained using the method of separation of variables. The solution is constructed using its general representation by two harmonic functions. Using the MATLAB software, numerical results and constructed graphs of the some boundary value problems are obtained.

Keywords: parabolic coordinates, separation of variables, elasticity, boundary, value problem, harmonic function

1. Introduction

In order to solve boundary value and boundary-contact problems in the areas with curvilinear border, it is purposeful to examine such problems in the relevant curvilinear coordinate system. Namely, the problems for the regions bounded by a circle or its parts are considered in the polar coordinate system [1–4], while the problems for the regions bounded by an ellipse or its parts or hyperbola are considered in the elliptic coordinate system [5–13], and the problems for the regions with parabolic boundaries are considered in the parabolic coordinate system [14–16]. The problems for the regions bounded by the circles with different centers and radiuses are considered in the bipolar coordinate system [17–19]. For that purpose, first the governing differential equations are expressed in terms of the relevant curvilinear coordinates. Then a number of important problems involving the relevant curvilinear coordinates are solved.

The chapter consists of five paragraphs.

Many problems are very easily cast in terms of parabolic coordinates. To this end, first the governing differential equations discussed in present chapter are expressed in terms of parabolic coordinates; then two concrete (test) problems involving parabolic coordinates are solved.

The second section, following the Introduction, gives the equilibrium equations and Hooke's law written down in the parabolic coordinate system and the setting of boundary value problems in the parabolic coordinate system.

Section 3 considers the method used to solve internal and external boundary value problems of elasticity for a homogeneous isotropic body bounded by parabolic curves.

IntechOpen

Section 4 solves the concrete problems, gains the numerical results, and constructs the relevant graphs.

Section 5 is a conclusion.

2. Problems statement

2.1 Equilibrium equations and Hooke's law in parabolic coordinates

It is known that elastic equilibrium of an isotropic homogeneous elastic body free of volume forces is described by the following differential equation [20]:

$$(\lambda + 2\mu)\mathrm{grad}\,\mathrm{div}\,\vec{U} - \mu\,\mathrm{rot}\,\mathrm{rot}\,\vec{U} = 0 \tag{1}$$

where $\lambda = E\nu/[(1+\nu)(1-2\nu)]$, $\mu = E/[2(1-\nu)]$ are elastic Lamé constants; ν is the Poisson's ratio; E is the modulus of elasticity; and $\vec{U}$ is a displacement vector.

By projecting Eq. (1) onto the tangent lines of the curves of the parabolic coordinate system (see Appendix A), we obtain the system of equilibrium equations in the parabolic coordinates.

In the parabolic coordinate system, the equilibrium equations with respect to the function D, K, u, v and Hooke's law can be written as [20–22]:

$$\begin{aligned} &(\mathrm{a})\,D_{,\xi} - K_{,\eta} = 0, \qquad (\mathrm{c})\,\overline{u}_{,\xi} + \overline{\mathrm{v}}_{,\eta} = (\kappa - 2)/(\kappa\mu)\cdot h_0^2 D,\\ &(\mathrm{b})\,D_{,\eta} + K_{,\xi} = 0, \qquad (\mathrm{d})\,\overline{\mathrm{v}}_{,\xi} - \overline{u}_{,\eta} = 1/\mu \cdot h_0^2 K. \end{aligned} \tag{2}$$

$$\begin{aligned} \sigma_{\eta\eta} &= h_0^{-1}\left[\lambda\overline{u}_{,\xi} + (\lambda + 2\mu)\overline{\mathrm{v}}_{,\eta} + \left[(\lambda+\mu) - \mu h_0^{-2}\right](\xi\overline{u} + \eta\overline{\mathrm{v}})\right],\\ \sigma_{\xi\xi} &= h_0^{-1}\left[(\lambda + 2\mu)\overline{u}_{,\xi} + \lambda\overline{\mathrm{v}}_{,\eta} + \left[(\lambda+\mu) + \mu h_0^{-2}\right](\xi\overline{u} + \eta\overline{\mathrm{v}})\right],\\ \tau_{\xi\eta} &= \mu h_0^{-1}\left[\left(\mathrm{v}_{,\xi} + u_{,\eta}\right) - h_0^{-2}(\xi\overline{\mathrm{v}} + \eta\overline{u})\right], \end{aligned} \tag{3}$$

where $\kappa = 4(1-\nu)$, $\overline{u} = hu/c^2$, $\overline{\mathrm{v}} = h\mathrm{v}/c^2$, $h_0 = \sqrt{\xi^2 + \eta^2}$, $h = h_\xi = h_\eta = c\sqrt{\xi^2+\eta^2}$ are Lamé coefficients (see Appendix A), u, v are the components of the displacement vector $\vec{U}$ along the tangents of η, ξ curved lines, and c is the scale factor (see Appendix A). And in the present paper, we take $c = 1$, $(\kappa - 2)/(\kappa\mu)\cdot D$ is the divergence of the displacement vector, K/μ is the rotor component of the displacement vector; $\sigma_{\xi\xi}$, $\sigma_{\eta\eta}$ and $\tau_{\xi\eta} = \tau_{\eta\xi}$ are normal and tangential stresses; and sub-indexes $()_{,\xi}$ and $()_{,\eta}$ denotes partial derivatives with relevant coordinates, for example, $K_{,\xi} = \frac{\partial K}{\partial \xi}$.

2.2 Boundary conditions

In the parabolic system of coordinates ξ, η $(-\infty < \xi < \infty,\ 0 \le \eta < \infty)$, exact solutions of two-dimensional static boundary value problems of elasticity are constructed for homogeneous isotropic bodies occupying domains bounded by coordinate lines of the parabolic coordinate system (see Appendix A).

The elastic body occupies the following domain (see **Figures 1** and **2**):

$$(\mathrm{a})\ D_1 = \{0 < \xi < \xi_1,\ 0 < \eta < \eta_1\},\ (\mathrm{b})\ D = \{-\xi_1 < \xi < \xi_1,\ 0 < \eta < \eta_1\}, \tag{4}$$

$$(\mathrm{a})\ \Omega_1 = \{0 < \xi < \xi_1,\ \eta_1 < \eta < \infty\},\ (\mathrm{b})\ \Omega = \{-\xi_1 \le \xi < \xi_1,\ \eta_1 \le \eta < \infty\}. \tag{5}$$

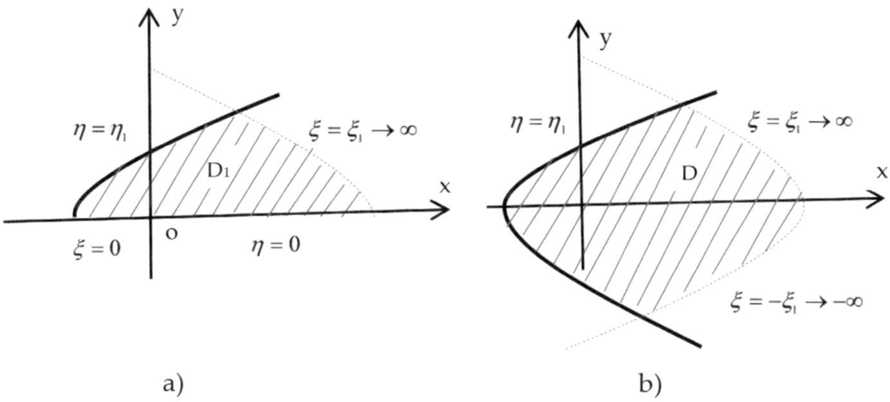

Figure 1.
(a) $D_1=\{0<\xi<\xi_1,\ 0<\eta<\eta_1\}$ *domain bounded by parabolic curve* $\eta=\eta_1$ *and line* y = 0 *and (b)* $D=\{-\xi_1<\xi<\xi_1,\ 0<\eta<\eta_1\}$ *domain bounded by parabola* $\eta=\eta_1$.

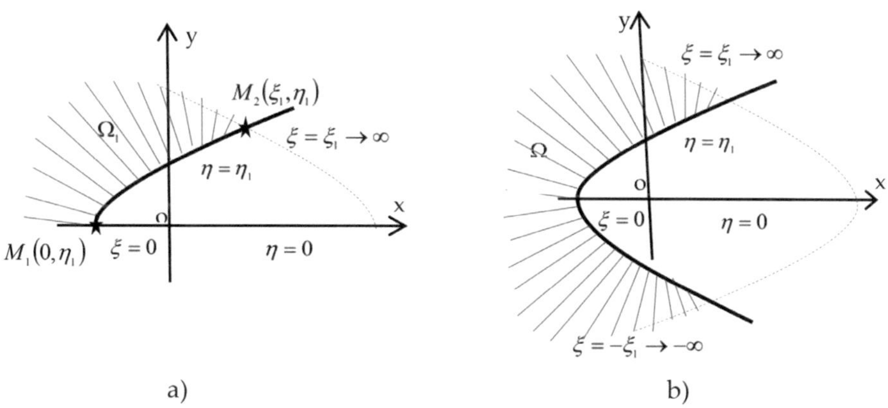

Figure 2.
Infinite domain (a) $\Omega_1=\{0<\xi<\xi_1,\ \eta_1<\eta<\infty\}$ *bounded by parabolic curve* $\eta=\eta_1$ *and line* y = 0 *and (b)* $\Omega=\{-\xi_1<\xi<\xi_1,\ \eta_1<\eta<\infty\}$ *bounded by parabola* $\eta=\eta_1$.

Boundary conditions that appear in the chapter have the following form:

$$\text{for}\quad \xi=\xi_1:\quad (\text{a})\,\sigma_{\xi\xi}=F^{1}{}_{(i)}(\eta),\quad \tau_{\xi\eta}=F_2^{(i)}(\eta)\quad\text{or}\quad(\text{b})\,u=G^{1}{}_{(i)}(\eta),\quad \text{v}=G_2^{(i)}(\eta), \tag{6}$$

$$\text{for}\quad \eta=\eta_1:\quad (\text{a})\,\sigma_{\eta\eta}=Q^{1}{}_{(i)}(\xi),\ \tau_{\xi\eta}=Q_2^{(i)}(\xi)\quad\text{or}\quad(\text{b})\,u=H^{1}{}_{(i)}(\xi),\quad \text{v}=H_2^{(i)}(\xi), \tag{7}$$

$$\text{for}\quad \xi=0:\quad (\text{a})\,\text{v}=0,\quad \sigma_{\xi\xi}=0\quad\text{or}\quad(\text{b})\,u=0,\quad \tau_{\xi\eta}=0, \tag{8}$$

$$\text{for}\,\eta=0:\quad (\text{a})\,u=0,\quad \sigma_{\eta\eta}=0,\quad\text{or}\quad(\text{b})\,\text{v}=0,\quad \tau_{\xi\eta}=0, \tag{9}$$

$$\text{for}\ \xi_1\to\pm\infty:\quad \sigma_{\eta\eta}\to 0,\quad \tau_{\xi\eta}\to 0,\quad u\to 0,\quad \text{v}\to 0. \tag{10}$$

$$\text{for}\ \eta\to\infty:\quad \sigma_{\eta\eta}\to 0,\quad \tau_{\xi\eta}\to 0,\quad u\to 0,\quad \text{v}\to 0, \tag{10a}$$

where F_i, Q_i $(i=1,2)$ with the first derivative and G_i, H_i with the first and second derivatives can be decomposed into the trigonometric absolute and uniform convergent Fourier series.

Boundary conditions on the linear parts $\xi = 0$ and $\eta = 0$ of the consideration area enable us to continue the solutions continuously (symmetrically or anti-symmetrically) in the domain, that is, the mirror reflection of the consideration area in a relationship $y = 0$ line (see **Figures 1b** and **2b**).

3. Solution of stated boundary value problems

In this section we will be considered internal and external problems for a homogeneous isotropic body bounded by parabolic curves.

3.1 Interior boundary value problems

Let us find the solution of problems (2), (3), (4a) (see **Figure 1a**), and (7)–(10) in class $C^2(D)$ (for D area shown in **Figure 1b**). The solution is presented by two harmonious φ_1 and φ_2 functions (see Appendix B). From formulas (B11)–(B13), after inserting $\alpha = \eta_1$ and making simple transformations, we will obtain:

$$\begin{aligned}\overline{u} &= -\left[\eta\left(\varphi_{1,\eta} - \varphi_{2,\xi}\right) + (\kappa - 1)\varphi\right]\xi + \left[\frac{\eta_1^2}{\eta}\left(\varphi_{1,\xi} + \varphi_{2,\eta}\right) - (\kappa - 1)\varphi_2\right]\eta,\\ \overline{\mathrm{v}} &= \left[\frac{\eta_1^2}{\eta}\left(\varphi_{1,\eta} - \varphi_{2,\xi}\right) + (\kappa - 1)\varphi_1\right]\eta + \left[\eta\left(\varphi_{1,\xi} + \varphi_{2,\eta}\right) - (\kappa - 1)\varphi_2\right]\xi;\end{aligned} \tag{11}$$

$$D = \frac{\kappa\mu}{h_0^2}\left[\left(\varphi_{1,\eta} - \varphi_{2,\xi}\right)\eta - \left(\varphi_{1,\xi} + \varphi_{2,\eta}\right)\xi\right], \quad K = \frac{\kappa\mu}{h_0^2}\left[\left(\varphi_{1,\eta} - \varphi_{2,\xi}\right)\xi + \left(\varphi_{1,\xi} + \varphi_{2,\eta}\right)\eta\right],$$

where

$$\frac{1}{h^2}\left(\varphi_{i,\xi\xi} + \varphi_{i,\eta\eta}\right) = 0, \quad i = 1, 2. \tag{12}$$

The stress tensor components can be written as

$$\begin{aligned}\frac{h_0^2}{2\mu}\sigma_{\eta\eta} = &-\left[\frac{\eta_1^2}{\eta}\left(\varphi_{1,\xi\xi} + \varphi_{2,\xi\eta}\right) - \frac{\kappa}{2}\varphi_{1,\eta} - \frac{\kappa - 2}{2}\varphi_{2,\xi}\right]\eta\\ &+ \left[\eta\left(\varphi_{1,\xi\eta} - \varphi_{2,\eta\eta}\right) + \frac{\kappa - 2}{2}\varphi_{1,\xi} - \frac{\kappa}{2}\varphi_{2,\eta}\right]\xi\\ &- \frac{\eta_1^2 - \eta}{\xi^2 + \eta^2}\left[\left(\varphi_{1,\eta} - \varphi_{2,\xi}\right)\eta - \left(\varphi_{1,\xi} + \varphi_{2,\eta}\right)\xi\right],\\ \frac{h_0^2}{2\mu}\tau_{\xi\eta} = &\left[\frac{\eta_1^2}{\eta}\left(\varphi_{1,\xi\eta} - \varphi_{2,\xi\xi}\right) + \frac{\kappa - 2}{2}\varphi_{1,\xi} - \frac{\kappa}{2}\varphi_{2,\eta}\right]\eta\\ &+ \left[\eta\left(\varphi_{1,\xi\xi} + \varphi_{2,\xi\eta}\right) - \frac{\kappa}{2}\varphi_{1,\eta} - \frac{\kappa - 2}{2}\varphi_{2,\xi}\right]\xi\\ &- \frac{\eta_1^2 - \eta}{\xi^2 + \eta^2}\left[\left(\varphi_{1,\eta} - \varphi_{2,\xi}\right)\xi + \left(\varphi_{1,\xi} + \varphi_{2,\eta}\right)\eta\right],\\ \frac{h_0^2}{2\mu}\sigma_{\xi\xi} = &\left[\frac{\eta_1^2}{\eta}\left(\varphi_{1,\xi\xi} + \varphi_{2,\xi\eta}\right) - \frac{\kappa - 4}{2}\varphi_{1,\eta} - \frac{\kappa + 2}{2}\varphi_{2,\xi}\right]\eta\\ &- \left[\eta\left(\varphi_{1,\xi\eta} - \varphi_{2,\xi\xi}\right) + \frac{\kappa + 2}{2}\varphi_{1,\xi} - \frac{\kappa - 4}{2}\varphi_{2,\eta}\right]\xi\\ &+ \frac{\eta_1^2 - \eta}{\xi^2 + \eta^2}\left[\left(\varphi_{1,\eta} - \varphi_{2,\xi}\right)\eta - \left(\varphi_{1,\xi} + \varphi_{2,\eta}\right)\xi\right].\end{aligned} \tag{13}$$

From (12) by the separation of variables method, we obtain (see Appendix A)

$$\varphi_i = \sum_{n=1}^{\infty} \varphi_{in}, \quad i = 1, 2, \tag{14}$$

where

$$\varphi_{1n} == A_{1n} \cosh(n\eta) \cos(n\xi), \qquad \varphi_{2n} == A_{2n} \sinh(n\eta) \sin(n\xi)$$

or

$$\varphi_{1n} == A_{1n} \sinh(n\eta) \sin(n\xi), \qquad \varphi_{2n} == A_{2n} \cosh(n\eta) \cos(n\xi).$$

For $n = 0$: $\varphi_{10} = A_{10} + a_{02}\xi + +a_{03}\eta + a_{04}\xi\eta$, $\varphi_{20} = A_{20} + b_{02}\xi + b_{03}\eta + b_{04}\xi\eta$, where $A_{10}, a_{02}, \ldots, b_{04}$ are constant coefficients. When $n = 0$ and $0 < \xi < \xi_1$, then the terms ξ, η and $\xi\eta$ will not be contained in φ_{10} and φ_{20}. If the foregoing solutions are presented in expressions of φ_{10} and φ_{20}, then it would be impossible on $\xi = \xi_1$ to satisfy the boundary conditions, and grad $\varphi_{i0} = (\varphi_{i0,\xi} + \varphi_{i0,\eta})/h$ $(i = 1, 2)$ will not be bounded in the point $M(0, 0)$.

Provision. We are introducing the following assumptions:

1. ξ_1 is a sufficiently great positive number (see Appendix C).

2. The boundary conditions given on $\eta = \eta_1$, i.e., stresses or displacements equal zero at interval $\tilde{\xi}_1 < \xi < \xi_1$.

3. When stresses are given on $\eta = \eta_1$, the main vector and main moment equal zero.

It is clear that

$$D = \kappa(\sigma_{\xi\xi} + \sigma_{\eta\eta})/4, \sigma_{\xi\xi} = 4D/\kappa - \sigma_{\eta\eta}.$$

By ultimately opening expressions $\sigma_{\eta\eta}$ and $\tau_{\xi\eta}$ (in details), we can demonstrate that at point $M(0, 0)$, $\sigma_{\eta\eta}$ and $\tau_{\xi\eta}$ (and naturally, $\sigma_{\xi\xi}$, too) are determined, i.e., they are finite.

When at $\eta = \eta_1$ $\overline{u}$ and $\overline{v}$ are given, then it is expedient to take instead of them as their equivalent the following expressions:

$$\begin{aligned} \frac{1}{h_0^2}(\overline{u} \cdot \eta_1 + \overline{v} \cdot \xi) &= \eta_1(\varphi_{1,\xi} + \varphi_{2,\eta}) - (\kappa - 1)\varphi_2, \\ \frac{1}{h_0^2}(\overline{u} \cdot \xi - \overline{v} \cdot \eta_1) &= \eta_1(\varphi_{1,\eta} - \varphi_{2,\xi}) + (\kappa - 1)\varphi_1, \end{aligned} \tag{15}$$

and if at $\eta = \eta_1$ $\frac{h_0^2}{2\mu}\sigma_{\eta\eta}$ and $\frac{h_0^2}{2\mu}\sigma_{\xi\eta}$ are given, then instead of them we have to take their equivalent following expressions:

$$\begin{aligned} \frac{1}{2\mu}(\sigma_{\eta\eta} \cdot \eta_1 - \sigma_{\xi\eta} \cdot \xi) &= -\eta_1(\varphi_{1,\xi\xi} + \varphi_{2,\xi\eta}) - \frac{\kappa}{2}\varphi_{1,\eta} - \frac{\kappa - 2}{2}\varphi_{2,\xi}, \\ \frac{1}{2\mu}(\sigma_{\eta\eta} \cdot \xi + \sigma_{\xi\eta} \cdot \eta_1) &= \eta_1(\varphi_{1,\xi\eta} - \varphi_{2,\xi\xi}) + \frac{\kappa - 2}{2}\varphi_{1,\xi} - \frac{\kappa}{2}\varphi_{2,\eta}. \end{aligned} \tag{16}$$

Considering the homogeneous boundary conditions of the concrete problem, we will insert φ_1 and φ_2 functions selected from the (14) in the right sides of (15) or

(16), and we will expand the left sides in the Fourier series. In both sides expressions which are with identical combinations of trigonometric functions will equate to each other and will receive the infinite system of linear algebraic equations to unknown coefficients A_{1n} and A_{2n} of harmonic functions, with its main matrix having a block-diagonal form. The dimension of each block is 2×2, and determinant is not equal to zero, but in infinite the determinant of block strives to the finite number different to zero.

It is very easy to establish the convergence of (11) and (13) functional series on the area $\overline{D} = \{-\xi_1 \le \xi \le \xi_1,\ 0 \le \eta \le \eta_1\}$ by construction of the corresponding uniform convergent numerical majorizing series. So we have the following:

Proposal 1. The functional series corresponding to (11) and (13) are absolute and uniform by convergent series on the area $\overline{D} = \{-\xi_1 \le \xi \le \xi_1,\ 0 \le \eta \le \eta_1\}$.

3.2 Exterior boundary value problems

We have to find the solution of problems (2), (3), (5a) (see **Figure 2a**), (7), (8), (10), and (10′), which belongs to the class $C^2(\Omega)$ (see region Ω on **Figure 2b**). The solution is constructed using its general representation by harmonic functions φ_1, φ_2 (see Appendix B). From formulas (B11)–(B13), following inserting $\alpha = \eta_1$ and simple transformations, we obtain the following expressions:

$$\overline{u} = -\left[(\varphi_{1,\xi} + \varphi_{2,\eta})\eta_1 + (\varphi_{1,\eta} - \varphi_{2,\xi})\xi\right](\eta - \eta_1) - \left[(\kappa - 1)\varphi_1 + \varphi_{3,\eta}\right]\xi - \left[(\kappa - 1)\varphi_2 - \varphi_{3,\xi}\right]\eta,$$

$$\overline{v} = \left[(\varphi_{1,\xi} + \varphi_{2,\eta})\xi - (\varphi_{1,\eta} - \varphi_{2,\xi})\eta_1\right](\eta - \eta_1) + \left[(\kappa - 1)\varphi_1 + \varphi_{3,\eta}\right]\eta - \left[(\kappa - 1)\varphi_2 - \varphi_{3,\xi}\right]\xi, \tag{17}$$

$$D = \frac{\kappa\mu}{h_0^2}\left[(\varphi_{1,\eta} - \varphi_{2,\xi})\eta - (\varphi_{1,\xi} + \varphi_{2,\eta})\xi\right],\ K = \frac{\kappa\mu}{h_0^2}\left[(\varphi_{1,\eta} - \varphi_{2,\xi})\xi + (\varphi_{1,\xi} + \varphi_{2,\eta})\eta\right],$$

where

$$\frac{1}{h^2}\left(\varphi_{i,\xi\xi} + \varphi_{i,\eta\eta}\right) = 0, \quad i = 1, 2, 3. \tag{18}$$

The stress tensor components can be written as:

$$\begin{aligned}\frac{h_0^2}{2\mu}\sigma_{\eta\eta} = {} & \left[(\varphi_{1,\xi\xi} + \varphi_{2,\xi\eta})\eta_1 + (\varphi_{1,\xi\eta} - \varphi_{2,\xi\xi})\xi\right](\eta - \eta_0) \\ & + \left(\frac{\kappa}{2}\varphi_{1,\eta} + \frac{\kappa - 2}{2}\varphi_{2,\xi} - \varphi_{3,\xi\xi}\right)\eta \\ & + \left(\frac{\kappa - 2}{2}\varphi_{1,\xi} - \frac{\kappa}{2}\varphi_{2,\eta} + \varphi_{3,\xi\eta}\right)\xi \\ & + \frac{\eta^2 - \eta_1^2}{\xi^2 + \eta^2}\left[(\varphi_{1,\eta} - \varphi_{2,\xi})\eta - (\varphi_{1,\xi} + \varphi_{2,\eta})\xi\right],\end{aligned}$$

$$\begin{aligned}\frac{h_0^2}{2\mu}\tau_{\xi\eta} = {} & \left[(\varphi_{1,\xi\xi} + \varphi_{2,\xi\eta})\xi - (\varphi_{1,\xi\eta} - \varphi_{2,\xi\xi})\eta_1\right](\eta - \eta_1) - \left(\frac{\kappa}{2}\varphi_{1,\eta} + \frac{\kappa - 2}{2}\varphi_{2,\xi} - \varphi_{3,\xi\xi}\right)\xi \\ & + \left(\frac{\kappa - 2}{2}\varphi_{1,\xi} - \frac{\kappa}{2}\varphi_{2,\eta} + \varphi_{3,\xi\eta}\right)\eta + \frac{\eta^2 - \eta_1^2}{\xi^2 + \eta^2}\left[(\varphi_{1,\eta} - \varphi_{2,\xi})\xi - (\varphi_{1,\xi} + \varphi_{2,\eta})\eta\right],\end{aligned} \tag{19}$$

$$\frac{h_0^2}{2\mu}\sigma_{\xi\xi} = -\left[(\varphi_{1,\xi\xi} + \varphi_{2,\xi\eta})\eta_1 + (\varphi_{1,\xi\eta} - \varphi_{2,\xi\xi})\xi\right](\eta - \eta_0)$$
$$-\left(\frac{\kappa - 4}{2}\varphi_{1,\eta} + \frac{\kappa + 2}{2}\varphi_{2,\xi} - \varphi_{3,\xi\xi}\right)\eta - \left(\frac{\kappa + 2}{2}\varphi_{1,\xi} - \frac{\kappa - 2}{2}\varphi_{2,\eta} + \varphi_{3,\xi\eta}\right)\xi$$
$$-\frac{\eta^2 - \eta_1^2}{\xi^2 + \eta^2}\left[(\varphi_{1,\eta} - \varphi_{2,\xi})\eta - (\varphi_{1,\xi} + \varphi_{2,\eta})\xi\right]$$

If $\overline{u}$ and $\overline{v}$ are given for $\eta = \eta_1$, then we take $\varphi_3 = 0$, and when $\frac{h_0^2}{2\mu}\sigma_{\eta\eta}$ and $\frac{h_0^2}{2\mu}\sigma_{\xi\eta}$ is given for $\eta = \eta_1$, then $\varphi_3 = \frac{\kappa-2}{2}\int\varphi_2 d\xi$.

From (18), by the separation of variables method, we obtain

$$\varphi_i = \sum_{n=1}^{\infty}\varphi_{in}, \quad i = 1, 2, 3, \tag{20}$$

where

$$\varphi_{1n} = B_{1n}e^{-n\eta}\sin(n\xi), \qquad \varphi_{2n} = B_{2n}e^{-n\eta}\cos(n\xi), \qquad \varphi_{3n} = \frac{\kappa - 2}{2n}B_{2n}e^{-n\eta}\sin(n\xi)$$

or

$$\varphi_{1n} = B_{1n}e^{-n\eta}\cos(n\xi), \qquad \varphi_{2n} = B_{2n}e^{-n\eta}\sin(n\xi), \qquad \varphi_{3n} = -\frac{\kappa - 2}{2n}B_{2n}e^{-n\eta}\cos(n\xi).$$

When $n = 0$, then $\varphi_{10} = A_{10} + a_{02}\xi + a_{03}\eta + a_{04}\xi\eta$, $\varphi_{20} = A_{20} + b_{02}\xi + b_{03}\eta + b_{04}\xi\eta$, where $A_{10}, a_{02}, \ldots, b_{04}$ are constants. From limited of functions φ_{i0} $(i = 1, 2)$ in $\eta \to \infty$ and satisfying boundary condition for $\xi = \xi_1$, it implies that $a_{02} = 0$, $b_{02} = 0$, $a_{03} = 0$, $b_{03} = 0$, $a_{04} = 0$, $b_{04} = 0$. Therefore, $\varphi_{10} = 0$, $\varphi_{20} = A_{20}$ or $\varphi_{10} = A_{10}$, $\varphi_{20} = 0$.

Provision. As in the previous subsection we make the following assumptions:

- ξ_1 is a sufficiently large positive number (see Appendix C).
- At $\eta = \eta_1$ given boundary conditions, i.e., displacements or stresses on interval $\tilde{\xi}_1 < \xi < \xi_1$, will equal zero.
- When stresses are given on $\eta = \eta_1$, the main vector and main moment will equal zero.

When $\overline{u}$ and $\overline{v}$ are given at $\eta = \eta_1$, then instead of them, it is expedient to take the following expressions as their equivalent:

$$\frac{1}{h_0^2(\kappa - 1)}(\overline{u}\xi - \overline{v}\eta_1) = \varphi_1, \qquad -\frac{1}{h_0^2(\kappa - 1)}(\overline{u}\eta_1 + \overline{v}\xi) = \varphi_2, \tag{21}$$

and if at $\eta = \eta_1$ $\frac{h_0^2}{2\mu}\sigma_{\eta\eta}$ and $\frac{h_0^2}{2\mu}\sigma_{\xi\eta}$ are given, then instead of them we have to take the following expressions as their equivalent:

$$\begin{aligned} \frac{1}{2\mu}\left(\sigma_{\eta\eta}\cdot\eta_1 - \sigma_{\xi\eta}\cdot\xi\right) &= \frac{\kappa}{2}\varphi_{1,\eta}, \\ \frac{1}{2\mu}\left(\sigma_{\eta\eta}\cdot\xi + \sigma_{\xi\eta}\cdot\eta_1\right) &= \frac{\kappa - 2}{2}\varphi_{1,\xi} - \varphi_{2,\eta}. \end{aligned} \tag{22}$$

Just like that in the previous subsection, considering the homogeneous boundary conditions of the concrete problem, we will insert φ_1 and φ_2 functions selected from (20) in Eq. (21) or (22), and we will expand the left sides in the Fourier series. Both sides of the expressions, which show the identical combinations of trigonometric functions, will equate to each other and will receive the infinite system of linear algebraic equations to unknown coefficients A_{1n} and A_{2n} of harmonic functions, with its main matrix having a block-diagonal form. The dimension of each block is 2×2, and the determinant does not equate to zero, but in the infinity, the determinant of block tends to the finite number different from zero.

As in the previous subsection, we received the following:

Proposition 2. The functional series corresponding to (17) and (19) are absolute and a uniformly convergent series on region $\overline{\Omega} = \{-\xi_1 \le \xi \le \xi_1, \ \eta_1 \le \eta < \infty\}$.

4. Test problems

In this section we will be obtained numerical results of internal and external problems for a homogeneous isotropic body bounded by parabolic curves when normal stress distribution is applied to the parabolic border.

4.1 Internal problem

We will set and solve the concrete internal boundary value problem in stresses. Let us find the solution of equilibrium equation system (2) of the homogeneous isotropic body in the area $\Omega_1 = \{0 < \xi < \xi_1, \ 0 < \eta < \eta_1\}$ (see **Figure 1a**), which satisfies boundary conditions (7a), (8a), (9a), and (10).

From (14), (8a), and (9a)

$$\varphi_i = \sum_{n=1}^{\infty} \varphi_{in}, \quad i = 1, 2, \tag{23}$$

where $\varphi_{1n} == A_{1n} \sinh(n\eta) \sin(n\xi)$, $\varphi_{2n} == A_{2n} \cosh(n\eta) \cos(n\xi)$.

By inserting (23) in (11) and (13), we will receive the following expressions for the displacements:

$$\begin{aligned}
\overline{u} = &\sum_{n=1}^{\infty} \{ -[n\eta\xi \cosh(n\eta)(A_{1n} + A_{2n}) + (\kappa - 1)\xi \sinh(n\eta) A_{1n}] \sin(n\xi) \\
&+ [n\eta_1^2 \sinh(n\eta)(A_{1n} + A_{2n}) - (\kappa - 1)\eta \cosh(n\eta) A_{2n}] \cos(n\xi) \}, \\
\overline{\mathrm{v}} = &\sum_{n=1}^{\infty} \{ [n\eta_1^2 \cosh(n\eta)(A_{1n} + A_{2n}) + (\kappa - 1)\eta \sinh(n\eta) A_{1n}] \sin(n\xi) \\
&+ [n\eta\xi \sinh(n\eta)(A_{1n} + A_{2n}) - (\kappa - 1)\xi \cosh(n\eta) A_{2n}] \cos(n\xi) \},
\end{aligned} \tag{24}$$

but for the stresses the following:

$$\begin{aligned}
\frac{h_0^2}{2\mu}\sigma_{\eta\eta} = &\sum_{n=1}^{\infty} \left\{ \left[n^2\eta_1^2 \sinh(n\eta)(A_{1n} + A_{2n}) + n\eta \cosh(n\eta)\left(\frac{\kappa}{2}A_{1n} - \frac{\kappa - 2}{2}A_{2n}\right) \right] \sin(n\xi) \right. \\
&+ \left[n^2\eta\xi \cosh(n\eta)(A_{1n} + A_{2n}) + n\xi \sinh(n\eta)\left(\frac{\kappa - 2}{2}A_{1n} - \frac{\kappa}{2}A_{2n}\right) \right] \cos(n\xi) \\
&\left. - \frac{\eta_1^2 - \eta^2}{\xi^2 + \eta^2}[n\eta \cosh(n\eta)(A_{1n} + A_{2n}) \sin(n\xi) - n\xi \sinh(n\eta)(A_{1n} + A_{2n}) \cos(n\xi)] \right\},
\end{aligned}$$

$$\frac{h_0^2}{2\mu}\tau_{\xi\eta} = \sum_{n=1}^{\infty}\left\{\left[n^2\eta_1^2\cosh(n\eta)(A_{1n}+A_{2n}) + n\eta\sinh(n\eta)\left(\frac{\kappa-2}{2}A_{1n} - \frac{\kappa}{2}A_{2n}\right)\right]\cos(n\xi)\right.$$
$$-\left[n^2\eta\xi\sinh(n\eta)(A_{1n}+A_{2n}) + n\xi\cosh(n\eta)\left(\frac{\kappa}{2}A_{1n} - \frac{\kappa-2}{2}A_{2n}\right)\right]\sin(n\xi)$$
$$\left.-\frac{\eta_1^2-\eta^2}{\xi^2+\eta^2}[n\xi\cosh(n\eta)(A_{1n}+A_{2n})\sin(n\xi) + n\eta\sinh(n\eta)(A_{1n}+A_{2n})\cos(n\xi)]\right\}, \tag{25}$$

$$\frac{h_0^2}{2\mu}\sigma_{\xi\xi} = \sum_{n=1}^{\infty}\left\{-\left[n^2\eta_1^2\sinh(n\eta)(A_{1n}+A_{2n}) + n\eta\cosh(n\eta)\left(\frac{\kappa-4}{2}A_{1n} - \frac{\kappa+2}{2}A_{2n}\right)\right]\sin(n\xi)\right.$$
$$-\left[n^2\eta\xi\cosh(n\eta)(A_{1n}+A_{2n}) + n\xi\sinh(n\eta)\left(\frac{\kappa+2}{2}A_{1n} - \frac{\kappa-4}{2}A_{2n}\right)\right]\cos(n\xi)$$
$$\left.+\frac{\eta_1^2-\eta^2}{\xi^2+\eta^2}[n\eta\cosh(n\eta)(A_{1n}+A_{2n})\sin(n\xi) - n\xi\sinh(n\eta)(A_{1n}+A_{2n})\cos(n\xi)]\right\}.$$

We have to solve problem (2), (7a), (8a), and (9a) when $Q_1(\xi) = P$ and $Q_2(\xi) = 0$, i.e., at $\eta = \eta_1$ boundary the normal load $\frac{1}{2\mu}\sigma_{\eta\eta} = \frac{P}{h_0^2}$ is given, but tangent stress is equal to zero. From (16), and (23), we obtain the following equations:

$$\sum_{n=1}^{\infty}\left[n^2\eta_1\sinh(n\eta_1)(A_{1n}+A_{2n}) - n\cosh(n\eta_1)\left(\frac{\kappa}{2}A_{1n} - \frac{\kappa-2}{2}A_{2n}\right)\right]\sin(n\xi) = \frac{P\eta_1}{\xi^2+\eta_1^2},$$
$$\sum_{n=1}^{\infty}\left[n^2\eta_1\cosh(n\eta_1)(A_{1n}+A_{2n}) + n\sinh(n\eta_1)\left(\frac{\kappa-2}{2}A_{1n} - \frac{\kappa}{2}A_{2n}\right)\right]\cos(n\xi) = \frac{P\xi}{\xi^2+\eta_1^2}.$$

From here an infinite system of the linear algebraic equations with unknown A_{1n} and A_{2n} coefficients is obtained:

$$\left[\left(n^2\eta_1\sinh(n\eta_1) - n\frac{\kappa}{2}\cosh(n\eta_1)\right)A_{1n}\right.$$
$$\left.+\left(n^2\eta_1\sinh(n\eta_1) + n\frac{\kappa-2}{2}\cosh(n\eta_1)\right)A_{2n}\right] = \tilde{F}_{1n},$$
$$\left[\left(n^2\eta_1\cosh(n\eta_1) + n\frac{\kappa-2}{2}\sinh(n\eta_1)\right)A_{1n}\right. \tag{26}$$
$$\left.+\left(n^2\eta_1\cosh(n\eta_1) - n\frac{\kappa}{2}\sinh(n\eta_1)\right)A_{2n}\right] = \tilde{F}_{2n}, \quad n = 1, 2, \dots$$

where $\tilde{F}_{1n}$ and $\tilde{F}_{2n}$ are the coefficients of expansion into the Fourier series $f_1(\xi) = \sum_{n=1}^{\infty}\tilde{F}_{1n}\sin(n\xi)$ and $f_2(\xi) = \sum_{n=1}^{\infty}\tilde{F}_{2n}\cos(n\xi)$, respectively, $f_1(\xi) = \frac{P\eta_1}{\xi^2+\eta_1^2}$ and $f_2(\xi) = \frac{P\xi}{\xi^2+\eta_1^2}$ functions.

As seen, the main matrix of system (26) has a block-diagonal form, dimension of each block is 2×2. Thus, two equations with two A_{1n} and A_{2n} unknown values will be solved. After solving this system, we find A_{1n} and A_{2n} coefficients, and in putting them into formulas (24) and (25), we get displacements and stresses at any points of the body.

Numerical values of displacements and stresses are obtained at the points of the finite size region bounded by curved lines $\eta = \eta_1$ and $\xi = \xi_1$ (see **Figure 1a**), and relevant 3D graphs are drafted. The numerical results are obtained for the following data: $\nu = 0.3$, $E = 2 \times 10^6 \text{kg/cm}^2$, $P = -10\,\text{kg/cm}^2$, $0.1 \le \eta_1 \le 3$, $\xi_1 = 2 * \pi$, $\xi_1 = 4 * \pi$, and $\xi_1 = 6 * \pi$. Numerical calculations and the visual presentation are made by MATLAB software.

Figures 3 and **4** show the distribution of stresses and displacements in the region bounded by curved lines $\eta = \eta_1$ and $\xi = \xi_{1k} := \xi_1$ (see **Figure 1a**), when (7a), (8a), and (9a) boundary conditions are valid and normal stress is applied to the parabolic boundary. Following conditions (8a) and (9a), at points of the linear parts $\xi = 0$ and $\eta = 0$ of consideration area $\sigma_{\xi\xi}(0, \eta)$, $\sigma_{\eta\eta}(\xi, 0)$ stresses and $u(\xi, 0)$, $v(0, \eta)$ displacements equal zero which is seen in **Figures 3** and **4**.

4.2 External problem

We will set and solve the concrete external boundary value problem in stresses. Let us find the solution of equilibrium equation system (2) of the homogeneous

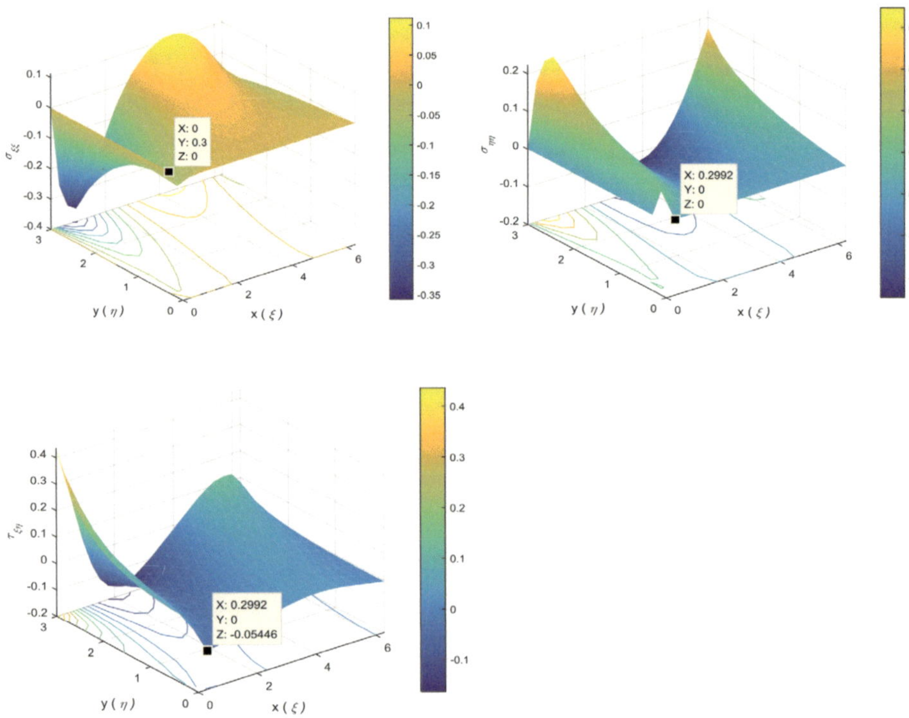

Figure 3.
Distribution of stresses in the region bounded by curved lines $\eta = \eta_1$ and $\xi = \xi_1$.

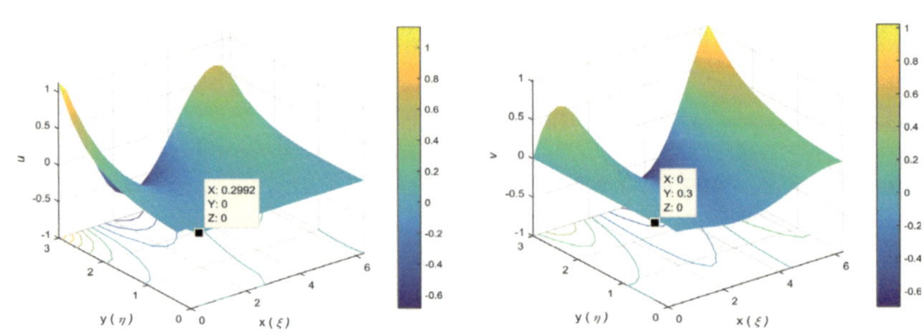

Figure 4.
Distribution of displacements in the region bounded by curved lines $\eta = \eta_1$ and $\xi = \xi_1$.

isotropic body in the region $\Omega_1 = \{0 < \xi < \xi_1,\ \eta_1 < \eta < \infty\}$, which satisfies the following boundary conditions: (7a), (8a), (10), and (10′).

From (20) and (8a)

$$\varphi_i = \sum_{n=1}^{\infty} \varphi_{in}, \quad i = 1, 2, 3, \tag{27}$$

where $\varphi_{1n} = B_{1n}e^{-n\eta}\sin(n\xi)$, $\varphi_{2n} = B_{2n}e^{-n\eta}\cos(n\xi)$, $\varphi_{3n} = \frac{\kappa-2}{2n}B_{2n}e^{-n\eta}\sin(n\xi)$.

By inserting (27) in (17) and (19), we will obtain the following expressions for displacements:

$$\begin{aligned}
\bar{u} &= \sum_{n=0}^{\infty}\{-ne^{-n\eta}[(B_{1n}-B_{2n})\eta_1\cos(n\xi) + (B_{1n}-B_{2n})\xi\sin(n\xi)](\eta-\eta_1) \\
&\quad -e^{-n\eta}[(\kappa-1)B_{1n} - (\kappa-2)B_{2n}]\xi\sin(n\xi) - \frac{\kappa}{2}e^{-n\eta}B_{2n}\eta\cos(n\xi)\Big\}, \\
\bar{\mathrm{v}} &= \sum_{n=1}^{\infty}\{ne^{-n\eta}[(B_{1n}-B_{2n})\xi\cos(n\xi) + (B_{1n}-B_{2n})\eta_1\sin(n\xi)](\eta-\eta_1) \\
&\quad +e^{-n\eta}[(\kappa-1)B_{1n} - (\kappa-2)B_{2n}]\eta\sin(n\xi) - \frac{\kappa}{2}e^{-n\eta}B_{2n}\xi\cos(n\xi)\Big\},
\end{aligned} \tag{28}$$

and for the stresses, we obtain the following formula:

$$\begin{aligned}
\frac{h_0^2}{2\mu}\sigma_{\eta\eta} &= \sum_{n=1}^{\infty}\Big\{-n^2e^{-n\eta}[(B_{1n}-B_{2n})\eta_1\sin(n\xi) + (B_{1n}-B_{2n})\xi\cos(n\xi)](\eta-\eta_1) \\
&\quad -ne^{-n\eta}\left[\frac{\kappa}{2}B_{1n}\eta\sin(n\xi) - \left(\frac{\kappa-2}{2}B_{1n} + B_{2n}\right)\xi\cos(n\xi)\right] \\
&\quad -\frac{\eta^2-\eta_1^2}{\xi^2+\eta^2}ne^{-n\eta}(B_{1n}-B_{2n})[\eta\sin(n\xi) + \xi\cos(n\xi)]\Big\}, \\
\frac{h_0^2}{2\mu}\tau_{\xi\eta} &= \sum_{n=1}^{\infty}\Big\{-n^2e^{-n\eta}[(B_{1n}-B_{2n})\xi\sin(n\xi) - (B_{1n}-B_{2n})\eta_1\cos(n\xi)](\eta-\eta_1) \\
&\quad -ne^{-n\eta}\left[\frac{\kappa}{2}B_{1n}\xi\sin(n\xi) - \left(\frac{\kappa-2}{2}B_{1n} + B_{2n}\right)\eta\cos(n\xi)\right], \\
&\quad -\frac{\eta^2-\eta_1^2}{\xi^2+\eta^2}ne^{-n\eta}(B_{1n}-B_{2n})[\xi\sin(n\xi) + \eta\cos(n\xi)]\Big\}, \\
\frac{h_0^2}{2\mu}\sigma_{\xi\xi} &= \sum_{n=1}^{\infty}\Big\{n^2e^{-n\eta}[(B_{1n}-B_{2n})\eta_1\sin(n\xi) + (B_{1n}-B_{2n})\xi\cos(n\xi)](\eta-\eta_1) \\
&\quad +ne^{-n\eta}\left[\left(\frac{\kappa-4}{2}B_{1n} + 2B_{2n}\right)\eta\sin(n\xi) + \frac{\kappa+2}{2}B_{1n}\xi\cos(n\xi)\right] \\
&\quad +\frac{\eta^2-\eta_1^2}{\xi^2+\eta^2}ne^{-n\eta}(B_{1n}-B_{2n})[\eta\sin(n\xi) + \xi\cos(n\xi)]\Big\}.
\end{aligned} \tag{29}$$

Next, we will obtain the numerical results of the following example.

We have to solve problem (2), (7a), and (8a), when $Q_1(\xi) = P$ and $Q_2(\xi) = 0$, i.e., at $\eta = \eta_1$ boundary the normal load $\frac{1}{2\mu}\sigma_{\eta\eta} = \frac{P}{h_0^2}$ is given, but tangent stress is equal to zero. From (22) and (27), we obtain the following equations:

$$\sum_{n=1}^{\infty} ne^{-n\eta_1}\frac{\kappa}{2}B_{1n}\sin(n\xi) = -\frac{P\eta_1}{\xi^2+\eta_1^2}, \quad \sum_{n=1}^{\infty} ne^{-n\eta_1}\left(\frac{\kappa-2}{2}B_{1n}+B_{2n}\right)\cos(n\xi) = \frac{P\xi}{\xi^2+\eta_1^2}.$$

Consequently, we obtain the infinite system of the linear algebraic equations with unknown B_{1n} and B_{2n} coefficients:

$$\sum_{n=1}^{\infty} ne^{-n\eta_1}\tfrac{\kappa}{2}B_{1n}\sin(n\xi) = -\sum_{n=1}^{\infty}\tilde{P}_{1n}\sin(n\xi), \quad \sum_{n=1}^{\infty} ne^{-n\eta_1}\left(\tfrac{\kappa-2}{2}B_{1n}+B_{2n}\right)\cos(n\xi) = \sum_{n=1}^{\infty}\tilde{P}_{2n}\cos(n\xi),$$

i.e.,

$$ne^{-n\eta_1}\frac{\kappa}{2}B_{1n} = -\tilde{P}_{1n}, \quad ne^{-n\eta_1}\left(\frac{\kappa-2}{2}B_{1n}+B_{2n}\right) = \tilde{P}_{2n}, \quad n = 1, 2, \ldots. \tag{30}$$

Hence,

$$B_{1n} = -\frac{2}{\kappa n}e^{n\eta_1}\tilde{P}_{1n}, \quad B_{2n} = \frac{e^{n\eta_1}}{n}\left(\tilde{P}_{2n}+\frac{\kappa-2}{\kappa}\tilde{P}_{1n}\right),$$

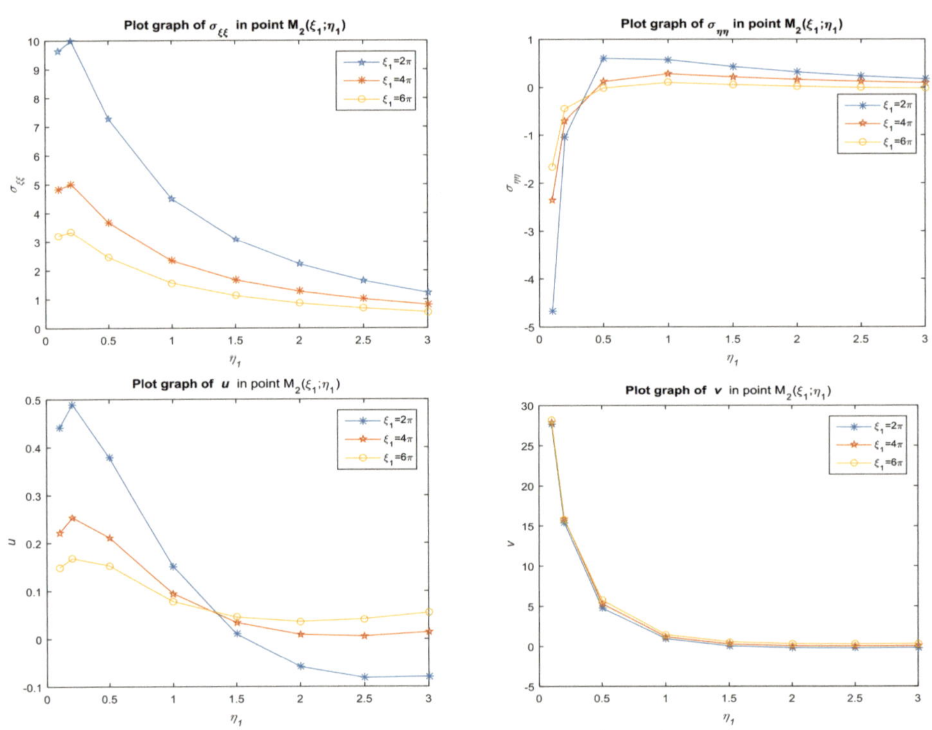

Figure 5.
Stresses and displacements at points $M_2(\xi_1, \eta_1)$ for $\xi_1 = 2*\pi$, $\xi_1 = 4*\pi$, and $\xi_1 = 6*\pi$, when $0.01 \le \eta_1 \le 3$.

where $\tilde{P}_{1n}$ and $\tilde{P}_{2n}$ are the coefficients of expansion into the Fourier series of functions $f_1(\xi) = -\frac{P\eta_1}{\xi^2+\eta_1^2}$ and $f_2(\xi) = \frac{P\xi}{\xi^2+\eta_1^2}$, respectively ($f_1(\xi)$, according to sinuses, and $f_2(\xi)$, according to cosines).

As it can be seen, the main matrix of system (30) has a block-diagonal form, and the dimension of each block is 2×2. Thus, two equations with two B_{1n} and B_{2n} unknown values will be solved. After solving this system, we find the values of B_{1n} and B_{2n} coefficients and put them into formulas (28) and (29) to get displacements and stresses at any points of the body.

Numerical results are obtained for some characteristic points of the body, in particular, $M_1(0,\eta_1)$, $M_2(\xi_1,\eta_1)$ points (see. **Figure 2a**), for the following data: $\nu = 0.3$, $E = 2*10^6\text{kg/cm}^2$, $P = -10\,\text{kg/cm}^2$, $0.01 \le \eta_1 \le 3$, $\xi_1 = 2*\pi$, $\xi_1 = 4*\pi$, and $\xi_1 = 6*\pi$.

The above-presented graphs (see **Figures 5** and **6**) show how displacements and stresses change at some characteristic points of body, namely, at points $M_1^{(j)}\left(0,\eta_1^{(j)}\right)$ and $M_2^{(j)}\left(\xi_1,\eta_1^{(j)}\right)$ $(j = 1, 2, \dots, 8)$, when $0.01 \le \eta_1 \le 3$ (see **Figure 7**).

From the presented results, we obtain the following:

- At points $M_1^{(j)}\left(0,\eta_1^{(j)}\right)$, $\max|u^t| < \max|u^n|$, $\quad v^t = v^n = 0$.

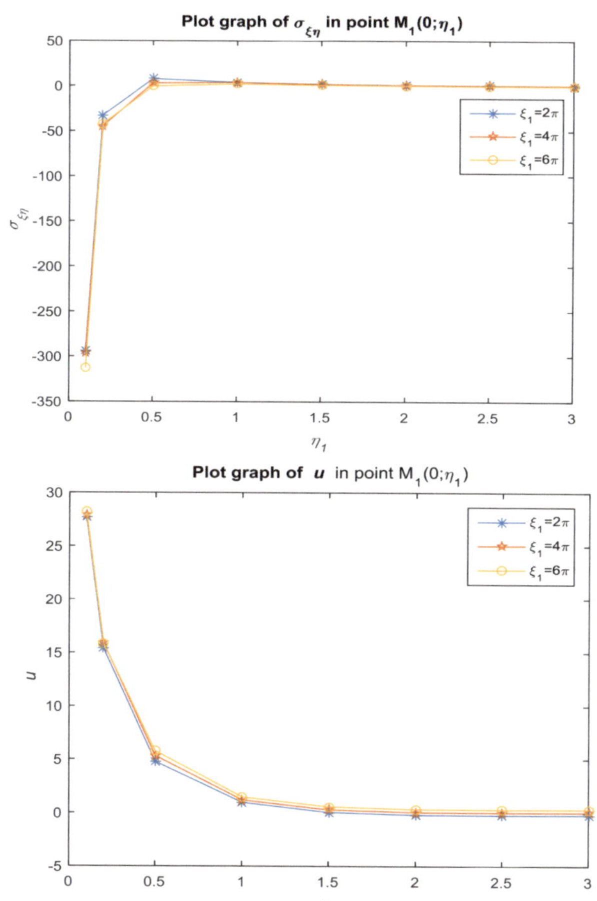

Figure 6.
Tangential stress and normal displacements at points $M_1(0,\eta_1)$ for $\xi_1 = 2*\pi$, $\xi_1 = 4*\pi$, and $\xi_1 = 6*\pi$, when $0.01 \le \eta_1 \le 3$.

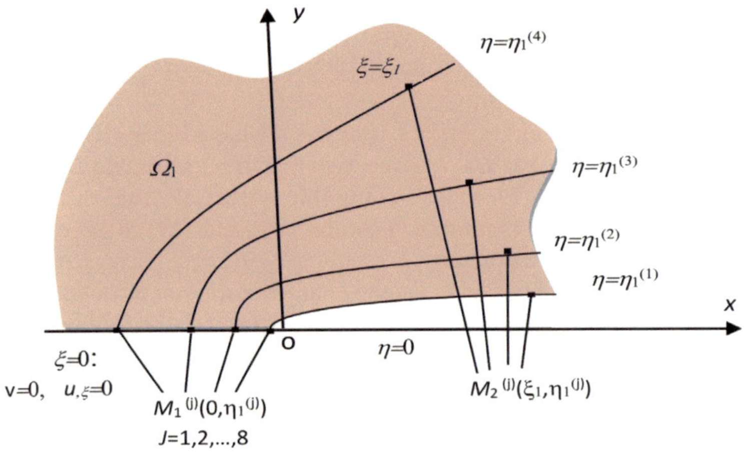

Figure 7.
Infinite region bounded by parabola marked with points, when obtaining the above-presented numerical results.

- At points $M_2^{(j)}\left(\xi_1,\eta_1^{(j)}\right)$, $\max\left|\sigma_{\xi\xi}^t\right| > \max\left|\sigma_{\xi\xi}^n\right|$, $\max|u^t| > \max|u^n|$, $\max|v^t| < \max|v^n|$.

- When $\xi_1 \to \infty$, then displacements and stresses tend to zero, that is, the boundary conditions (10) are satisfied.

- When $\eta_1 \to \infty$, then displacements and stresses tend to zero, that is, the boundary conditions (10′) are satisfied.

- When $\eta_1 \to 0$ (in this case there is a crack), then (a) at points $M_1^{(j)}\left(0,\eta_1^{(j)}\right)$ tangential stresses and normal displacements tend to ∞, but other components equal to zero. It can be seen from the boundary conditions (8a) (b) at points $M_2^{(j)}\left(\xi_1,\eta_1^{(j)}\right)$ that all components of the displacements and stresses tend to ∞.

Here superscript t and n denote the tangential and normal displacement or the stress, respectively.

5. Conclusion

The main results of this chapter can be formulated as follows:

- The equilibrium equations and Hooke's law are written in terms of parabolic coordinates.

- The solution of the equilibrium equations is obtained by the method of separation of variables. The solution is constructed using its general representation by harmonic functions.

- In parabolic coordinates, analytical solutions of 2D static boundary value problems for the elasticity are constructed for homogeneous isotropic finite and infinite bodies occupying domains bounded by coordinate lines of parabolic coordinate system.

- Two concrete internal and external boundary value problems in stresses are set and solved.

The bodies bounded by the parabola are common in practice, for example, in building, mechanical engineering, biology, medicine, etc., the study of the deformed state of such bodies is topical, and consequently, in my opinion, setting the problems considered in the chapter and the method of their solution is interesting in a practical view.

Notations

x, y	Cartesian coordinates
ξ, η	parabolic coordinates
E and v	modulus of elasticity and Poisson's ratio
λ, μ	elastic Lamé constants
$\vec{U}(u, \mathrm{v})$	displacement vector
$\sigma_{\xi\xi}, \sigma_{\eta\eta}, \tau_{\xi\eta} = \tau_{\eta\xi}$	normal and tangential stresses

Appendix

A. Some basic formulas in parabolic coordinates

In orthogonal parabolic coordinate system $\xi, \eta(-\infty < \xi < \infty,\ 0 \leq \eta < \infty$, see **Figure A1**) [23, 24]; we have

$$h_\xi = h_\eta = h = c\sqrt{\xi^2 + \eta^2}, \quad x = c\left(\xi^2 - \eta^2\right)/2, \quad y = c\xi\eta,$$

where h_ξ, h_η are Lame's coefficients of the system of parabolic coordinates, c is a scale coefficient, x, y are the Cartesian coordinates.

The coordinate axes are parabolas

$$y^2 = -2c\xi_0^2\left(x - c\xi_0^2/2\right), \quad \xi_0 = \text{const}, \quad y^2 = -2c\eta_0^2\left(x + c\eta_0^2/2\right), \quad \eta_0 = \text{const}.$$

Laplace's equation $\Delta f = 0$, where $f = f(\xi, \eta)$, in the parabolic coordinates has the form

$$\left(f_{,\xi\xi} + f_{,\eta\eta}\right)/c^2\left(\xi^2 + \eta^2\right) = 0.$$

We have to find solution of the equation in following form

$$f = X(\xi) \cdot E(\eta),$$

and then by separation of variables, we will receive

$$\frac{1}{c^2\left(\xi^2 + \eta^2\right)}\left[\frac{X''}{X} + \frac{E'}{E}\right] = 0.$$

From here

$$X'' + mX = 0, \qquad E'' - mE = 0,$$

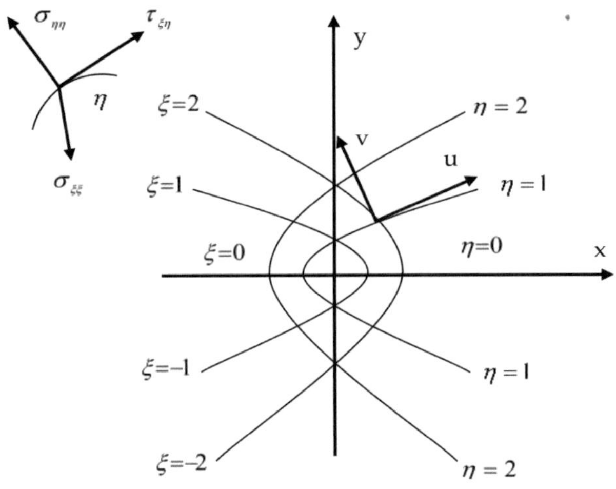

Figure A1.
Parabolic coordinate system.

where m is any constant, their solutions are [25]

$$X = C_1 \cos(m\xi) + C_2 \sin(m\xi), \quad E = C_3 e^{m\eta} + C_4 e^{-m\eta}$$
$$= C_3^* \cosh(m\eta) + C_4^* \sinh(m\eta).$$

So

$$f(\xi, \eta) = \left(C_3 e^{m\eta} + C_4 e^{-m\eta}\right)\left(C_1 \cos(m\xi) + C_2 \sin(m\xi)\right)$$

or

$$f(\xi, \eta) = \left(C_3^* \cosh(m\eta) + C_4^* \sinh(m\eta)\right)\left(C_1 \cos(m\xi) + C_2 \sin(m\xi)\right),$$

B. Solution of system of partial differential equations

We solve the system of partial differential equations (2).
We have introduced φ_1 harmonic function, and if we take

$$\text{(a)}\ D = \frac{\kappa\mu}{h_0^2}\left(\varphi_{1,\eta}\eta - \varphi_{1,\xi}\xi\right), \qquad \text{(b)}\ K = \frac{\kappa\mu}{h_0^2}\left(\varphi_{1,\eta}\xi + \varphi_{1,\xi}\eta\right), \tag{B1}$$

then Eqs. (2a) and (2b) will be satisfied identically, while Eqs. (2c) and (2d) will receive the following form:

$$\text{(a)}\ \overline{u}_{,\xi} + \overline{v}_{,\eta} = (\kappa - 2)\left(\varphi_{1,\eta}\eta - \varphi_{1,\xi}\xi\right), \quad \text{(b)}\ \overline{v}_{,\xi} - \overline{u}_{,\eta} = \kappa\left(\varphi_{1,\eta}\xi + \varphi_{1,\xi}\eta\right), \tag{B2}$$

$$\text{(a)}\ \overline{u}_{,\xi} + \overline{v}_{,\eta} = (\kappa - 2)\left(\varphi_{1,\eta}\eta - \varphi_{1,\xi}\xi\right), \quad \text{(b)}\ \left(\overline{v} - \kappa\varphi_1\eta\right)_{,\xi} = \left(\overline{u} + \kappa\varphi_1\xi\right)_{,\eta}. \tag{B3}$$

From equation (B3b) imply that exists such type harmonic function φ, for which fulfill the following

$$\overline{u} = \varphi_{,\xi} - \kappa\varphi_1\xi, \qquad \overline{v} = \varphi_{,\eta} + \kappa\varphi_1\eta. \tag{B4}$$

Considering (B4), from Equation (B3a), the following will be obtained:

$$h^2\Delta\varphi = \varphi_{,\xi\xi} + \varphi_{,\eta\eta} = \kappa\varphi_1 + \kappa\varphi_{1,\xi}\xi - \kappa\varphi_1 - \kappa\varphi_{1,\eta}\eta + (\kappa - 2)\left(\varphi_{1,\eta}\eta - \varphi_{1\xi}\xi\right)$$
$$= 2\left(\varphi_{1,\xi}\xi - \varphi_{1,\eta}\eta\right). \tag{B5}$$

General solution of the system (B2) can be written in the form $\overline{u} = \psi_1$, $\overline{v} = \psi_2$, where

$$\psi_{1,\xi} + \psi_{2,\eta} = 0, \qquad \psi_{2,\xi} - \psi_{1,\eta} = 0.$$

The full solution of equation system (B2) is written in the following form:

$$\overline{u} = \varphi_{,\xi} - \kappa\varphi_1\xi + \psi_1, \qquad \overline{v} = \varphi_{,\eta} + \kappa\varphi_1\eta + \psi_2, \tag{B6}$$

where φ is the partial solution of the (B5).
If we take $\kappa = \text{const}$, then

$$\varphi = \frac{\xi^2 - \eta^2}{2}\varphi_1,$$

and (B6) formula will receive the following form:

$$\overline{u} = \frac{\xi^2 - \eta^2}{2}\varphi_{1,\xi} - (\kappa - 1)\varphi_1\xi + \psi_1, \qquad \overline{v} = \frac{\xi^2 - \eta^2}{2}\varphi_{1,\eta} + (\kappa - 1)\varphi_1\eta + \psi_2.$$

From here

$$\overline{u} = \left(\frac{\xi^2 - \eta^2}{2}\varphi_{1,\xi} + \xi\eta\varphi_{1,\eta}\right) - \xi\eta\varphi_{1,\eta} - (\kappa - 1)\varphi_1\xi + \psi_1,$$

$$\overline{v} = \left(\frac{\xi^2 - \eta^2}{2}\varphi_{1,\eta} - \xi\eta\varphi_{1,\xi}\right) + \xi\eta\varphi_{1,\xi} + (\kappa - 1)\varphi_1\eta + \psi_2.$$

Without losing the generality, the expression in brackets can be taken as zero, because we already have in $\overline{u}$ and $\overline{v}$ of the solutions Laplacian (we mean ψ_1 and ψ_2). Therefore, the solutions of system (2) are given in the following form:

$$\begin{aligned} &\text{(a)}\ h_0^2 D = \kappa\mu\left(\varphi_{1,\eta}\eta - \varphi_{1,\xi}\xi\right), && \text{(b)}\ h_0^2 K = \kappa\mu\left(\varphi_{1,\eta}\xi + \varphi_{1,\xi}\eta\right), \\ &\text{(c)}\ \overline{u} = -\xi\eta\varphi_{1,\eta} - (\kappa - 1)\varphi_1\xi + \psi_1, && \text{(d)}\ \overline{v} = \xi\eta\varphi_{1,\xi} + (\kappa - 1)\varphi_1\eta + \psi_2. \end{aligned} \tag{B7}$$

Now we have to write down three versions of ψ_1 and ψ_2 function representation. In the first version

$$\psi_1 = \overline{\varphi}_{1,\eta} + \tilde{\varphi}_{1,\eta} + \varphi_{2,\eta}, \qquad \psi_2 = \overline{\varphi}_{1,\xi} + \tilde{\varphi}_{1,\xi} + \varphi_{2,\xi}, \tag{B8}$$

$\overline{\varphi}_1, \tilde{\varphi}_1, \varphi_2$ are harmonic functions; in addition, $\overline{\varphi}_1, \tilde{\varphi}_1$ are selected so that at $\eta = \alpha$, where $\alpha = \eta_1$ or $\alpha = \eta_2$, the following equations will be satisfied:

$$-\xi\eta\varphi_{1,\eta} - (\kappa - 1)\varphi_1\xi + \overline{\varphi}_{1,\eta} + \tilde{\varphi}_{1,\eta} = 0, \qquad \xi\eta\varphi_{1,\xi} + (\kappa - 1)\varphi_1\xi + \overline{\varphi}_{1,\xi} + \tilde{\varphi}_{1,\xi} = 0,$$

In the second version

$$\begin{aligned} \psi_1 &= -\alpha\left(\frac{\xi^2 - (\eta - \alpha)^2}{2}\varphi_{1,\xi} + \xi\eta\varphi_{1,\eta}\right) + \frac{\xi^2 - \eta^2}{2}\varphi_{2,\xi} + \xi\eta\varphi_{2,\eta}, \\ \psi_2 &= \alpha\left(\xi\eta\varphi_{1,\xi} - \frac{\xi^2 - (\eta - \alpha)^2}{2}\varphi_{1,\eta}\right) + \frac{\xi^2 - \eta^2}{2}\varphi_{2,\eta} - \xi\eta\varphi_{2,\xi}, \end{aligned} \tag{B9}$$

where φ_2 is the harmonic function.
In the third version

$$\psi_1 = -\alpha^2\left(\frac{\xi^2-\eta^2}{2}\varphi_{1,\xi} + \xi\eta\varphi_{1,\eta}\right) + \frac{\xi^2-\eta^2}{2}\varphi_{2,\xi} + \xi\eta\varphi_{2,\eta},$$
$$\psi_2 = \alpha^2\left(\xi\eta\varphi_{1,\xi} - \frac{\xi^2-\eta^2}{2}\varphi_{1,\eta}\right) + \frac{\xi^2-\eta^2}{2}\varphi_{2,\eta} - \xi\eta\varphi_{2,\xi}. \tag{B10}$$

Inserting (B8) in (B7c and d), we will get

$$\text{(a)}\,\overline{u} = -\xi\eta\varphi_{1,\eta} - (\kappa-1)\varphi_1\xi + \overline{\varphi}_{1,\eta} + \tilde{\varphi}_{1,\eta} + \varphi_{2,\eta},$$
$$\text{(b)}\,\overline{\mathrm{v}} = \xi\eta\varphi\xi + (\kappa-1)\varphi_1\xi + \overline{\varphi}_{1,\xi} + \tilde{\varphi}_{1,\xi} + \varphi_{2,\xi}. \tag{B11}$$

Inserting (B9) in (B7c and d), we will have

$$\text{(a)}\,\overline{u} = -\alpha\left(\frac{\xi^2-(\eta-\alpha)^2}{2}\varphi_{1,\xi} + \xi\eta\varphi_{1,\eta}\right) - \xi\eta\varphi_{1,\eta} - (\kappa-1)\varphi_1\xi + \frac{\xi^2-\eta^2}{2}\varphi_{2,\xi} + \xi\eta\varphi_{2,\eta},$$
$$\text{(b)}\,\overline{\mathrm{v}} = \alpha\left(\xi\eta\varphi_{1,\xi} - \frac{\xi^2-(\eta-\alpha)^2}{2}\varphi_{1,\eta}\right) + \xi\eta\varphi_{1,\xi} + (\kappa-1)\varphi_1\eta + \frac{\xi^2-\eta^2}{2}\varphi_{2,\eta} - \xi\eta\varphi_{2,\xi}. \tag{B12}$$

Inserting (B10) in (B7c and d), we will get

$$\text{(a)}\,\overline{u} = -\alpha^2\left(\frac{\xi^2-\eta^2}{2}\varphi_{1,\xi} + \xi\eta\varphi_{1,\eta}\right) - \xi\eta\varphi_{1,\eta} - (\kappa-1)\varphi_1\xi + \frac{\xi^2-\eta^2}{2}\varphi_{2,\xi} + \xi\eta\varphi_{2,\eta},$$
$$\text{(b)}\,\overline{\mathrm{v}} = \alpha^2\left(\xi\eta\varphi_{1,\xi} - \frac{\xi^2-\eta^2}{2}\varphi_{1,\eta}\right) + \xi\eta\varphi_{1,\xi} + (\kappa-1)\varphi_1\eta + \frac{\xi^2-\eta^2}{2}\varphi_{2,\eta} - \xi\eta\varphi_{2,\xi}. \tag{B13}$$

C. Finding of ξ_1

After the boundary value problem with relevant boundary conditions on $\xi = \xi_1 = \xi_{11}$ is solved, the following condition is examined: $F_{11}/F_{10} < \varepsilon$.

ε is a sufficiently small positive number given in advance ($\varepsilon = 0,001 - 0,0001$).

$$F_{11} = \left[\int_0^{\eta_1}\left(|\sigma_{\xi\xi}| + |\sigma_{\eta\eta}| + |\tau_{\xi\eta}|\right)hd\eta\right]_{\xi=\xi_1}, \quad F_{10} = \left[\int_0^{\eta_1}\left(|\sigma_{\xi\xi}| + |\sigma_{\eta\eta}| + |\tau_{\xi\eta}|\right)hd\eta\right]_{\xi=g\tilde{\xi}_1}.$$

g number will be selected so that on boundary $\eta = \eta_1$, point $M(g\tilde{\xi}_1, \eta_1)$ should correspond to the highest value of expression $\left[\sigma_{\eta\eta}(g\tilde{\xi}_1,\eta_1)\right]^2 + \left[\tau_{\xi\eta}(g\tilde{\xi}_1,\eta_1)\right]^2$ (when stresses are given) or to the highest value of expression $\left[\overline{u}(g\tilde{\xi}_1,\eta_1)\right]^2 + \left[\overline{\mathrm{v}}(g\tilde{\xi}_1,\eta_1)\right]^2$ (when displacements are given).

If condition $F_{11}/F_{10} < \varepsilon$ is not valid for $\xi_1 = \xi_{11}$, the same problem will be solved at the beginning, but $\xi_1 = \xi_{12}$ will be used instead of $\xi_1 = \xi_{11}$. In addition, $\xi_{12} > \xi_{11}$. Then, if condition $F_{12}/F_{10} < \varepsilon$ is not still valid, we will continue with the boundary problem, where $\xi_1 = \xi_{13}$; besides, $\xi_{13} > \xi_{12} > \xi_{11}$, and we will examine condition $F_{13}/F_{10} < \varepsilon$. The process will be over at the kth stage, if condition $F_{1k}/F_{10} < \varepsilon$ is valid.

Finding such $\xi_1 = \xi_{1k}$***, for which*** $F_{1k}/F_{10} < \varepsilon$.

Distance l between surfaces $\xi = \xi_1$ and $\xi = \tilde{\xi}_1$, which gives the guarantee for condition $F_{1k}/F_{10} < \varepsilon$ to be valid in the parabolic coordinate system, will be taken along the axis of the parabola , and the following expression will be obtained:

$$\xi_1 = \sqrt{l/c + \tilde{\xi}_1^2}.$$

By relying on the known solutions of the relevant plain problems of elasticity, it is purposeful to admit that $l/c = 4, 5, 6, \ldots,$ which allows finding ξ_1 from the relevant equation. Let us note that when $l/c = 4$, we will denote value ξ_1 by ξ_{11}, when $l/c = 5$; by ξ_{12}, when $l/c = 6$; by ξ_{13}, etc. If after selecting $\xi_1 = \xi_{1k}$, inequality $F_{1k}/F_{10} < \varepsilon$ is valid; in order to check the righteousness of the selection, it is necessary to once again make sure that, together with condition $F_{1k}/F_{10} < \varepsilon$, condition $\varepsilon > F_{1k}/F_{10} > F_{1k+1}/F_{10} > F_{1k+2}/F_{10} > \ldots$ is valid, too.

Author details

Natela Zirakashvili
I. Vekua Institute of Applied Mathematics of Iv. Javakhishvili Tbilisi State University, Tbilisi, Georgia

*Address all correspondence to: natzira@yahoo.com

References

[1] Muskhelishvili NI. Some Basic Problems of the Mathematical Theory of Elasticity. Groningen: Noordhoff; 1953. p. 731

[2] Khomasuridze N. Thermoelastic equilibrium of bodies in generalized cylindrical coordinates. Georgian Mathematical Journal. 1998;**5**:521-544

[3] Khomasuridze N, Zirakashvili N. Strain control of cracked elastic bodies by means of boundary condition variation. In: Proceedings of International Conference on Architecture and Construction – Contemporary Problems, 30 September–3 October 2010; Yerevan, Jermuk, Armenia. 2010. pp. 158-163. (in Russian)

[4] Zirakashvili N. Application of the boundary element method to the solution of the problem of distribution of stresses in an elastic body with a circular hole whose interior surface contains radial cracks. Proceedings of A. Razmadze Mathematical Institute. 2006;**141**:139-147

[5] Tang R, Wang Y. On the problem of crack system with an elliptic hole. Acta Mechanica Sinica. 1986;**2**(1):47-53

[6] Zirakashvili N. The numerical solution of boundary value problems for an elastic body with an elliptic hole and linear cracks. Journal of Engineering Mathematics. 2009;**65**(2):111-123

[7] Shestopalov Y, Kotik N. Approximate decomposition for the solution of boundary value problems for elliptic systems arising in mathematical models of layered structures. In: Progress in Electromagnetic Research Symposium; 26–29 March; Cambridge, MA. 2006. pp. 514-518

[8] Zirakashvili N. Some boundary value problems of elasticity for semi-ellipses. Proceedings of I. Vekua Institute of Applied Mathematics. 2002;**52**:49-55

[9] Zirakashvili N. Study of deflected mode of ellipse and ellipse weakened with crack. ZAMM. 2017;**97**(8):932-945

[10] Zirakashvili N. Boundary value problems of elasticity for semi-ellipse with non-homogeneous boundary conditions at the segment between focuses. Bulletin of TICMI. 2017;**21**(2): 95-116

[11] Zirakashvili N. Analytical solutions of boundary-value problems of elasticity for confocal elliptic ring and its parts. Journal of the Brazilian Society of Mechanical Sciences and Engineering. 2018;**40**(398):1-19

[12] Zirakashvili N. Analytical solutions of some internal boundary value problems of elasticity for domains with hyperbolic boundaries. Mathematics and Mechanics of Solids. 2019;**24**(6): 1726-1748

[13] Zirakashvili N. Study of stress-strain state of elastic body with hyperbolic notch. Zeitschrift für Angewandte Mathematik und Physik. 2019;**70**(87): 1-19

[14] Zirakashvili N. Analytical solution of interior boundary value problems of elasticity for the domain bounded by the parabola. Bulletin of TICMI. 2016;**20**(1): 3-24

[15] Zirakashvili N. Exact solution of some exterior boundary value problems of elasticity in parabolic coordinates. Mathematics and Mechanics of Solids. 2018;**23**(6):929-943

[16] Zappalorto M, Lazzarin P, Yates JR. Elastic stress distributions for hyperbolic and parabolic notches in round shafts under torsion and uniform anti-plane shear loadings. International

Journal of Solids and Structures. 2008; **45**:4879-4901

[17] Jeffery GB. Plane stress and plane strain in bipolar coordinates. Philosophical Transactions of the Royal Society of London. 1921;**221**:265-293

[18] Ufland YS. Bipolar Coordinates in Elasticity. Moscow-Leningrad: Gostehteoretizdat; 1950. p. 232. (in Russian)

[19] Khomasuridze N. Solution of some elasticity boundary value problems in bipolar coordinates. Acta Mechanica. 2007;**189**:207-224

[20] Novozhilov VV. Elasticity Theory. Sudpromgiz: Leningrad; 1958. p. 371. (in Russian)

[21] Khomasuridze N. Representation of solutions of some boundary value problems of elasticity by a sum of the solutions of other boundary value problems. Georgian Mathematical Journal. 2005;**10**(2):257-270

[22] Filonenko-Borodich M. Theory of Elasticity. Moscow: Gos. Izd. Phiz. -Mat. Lit.; 1959. p. 364. (in Russian)

[23] Lebedev NN. Special Functions and their Applications. Gosizdat of Phys.-Mat. Moscow-Leningrad: Literature; 1963. p. 359. (in Russian)

[24] Bermant AF. Mapping. Curvilinear Coordinates. Transformations. Green's Formula. Moscow: Gosizdat Fizmatgiz; 1958. p. 308. (in Russian)

[25] Kamke E. Handbook of Ordinary Differential Equations. Moscow: Nauka; 1971. p. 584. (in Russian)

Chapter 12

Boundary Element Mathematical Modelling and Boundary Element Numerical Techniques for Optimization of Micropolar Thermoviscoelastic Problems in Solid Deformable Bodies

Mohamed Abdelsabour Fahmy

Abstract

The main objective of this chapter is to introduce a new theory called three-temperature nonlinear generalized micropolar thermoviscoelasticity. Because of strong nonlinearity of simulation and optimization problems associated with this theory, the numerical solutions for problems related with the proposed theory are always very difficult and require the development of new numerical techniques. So, we propose a new boundary element technique for simulation and optimization of such problems based on genetic algorithm (GA), free form deformation (FFD) method and nonuniform rational B-spline curve (NURBS) as the shape optimization technique. In the formulation of the considered problem, the profiles of the considered objects are determined by FFD method, where the FFD control points positions are treated as genes, and then the chromosomes profiles are defined with the genes sequence. The population is founded by a number of individuals (chromosomes), where the objective functions of individuals are determined by the boundary element method (BEM). The numerical results verify the validity and accuracy of our proposed technique.

Keywords: boundary element modeling, simulation, optimization, three-temperature, nonlinear generalized micropolar thermoviscoelasticity, solid deformable bodies

1. Introduction

The classical thermo-elasticity (CTE) theory which was introduced by Duhamel [1] and Neumann [2] characterized the strain-temperature gradients equations in an elastic body, but it has two shortcomings contrary to physical observations: First, the heat conduction equation of this theory does not include any elastic terms. Second, the heat conduction equation is of a parabolic type predicting infinite speeds of thermal waves. Biot [3] developed the classical coupled thermo-elasticity (CCTE) theory to overcome the first shortcoming in CTE. However, both theories share the second shortcoming. So, several generalized thermoelasticity theories that predict finite speeds of propagation for heat waves have been developed such as extended thermo-elasticity (ETE) theory of Lord and Shulman [4], temperature-rate-dependent thermo-elasticity (TRDTE) theory of Green and Lindsay [5], three linear generalized thermoelasticity theories (type I, II and III) of Green and Naghdi [6, 7], low-temperature thermoelasticity (LTTE) model of Hetnarski and Ignaczak [8], the dual phase-lag (DPL) heat conduction equation of Tzou [9, 10] which has been developed taking into consideration the phonons-electrons interactions to obtain dual phase-lag thermoelasticity (DPLTE) [11, 12], and three-phase-lag thermoelasticity (TPLTE) model of Choudhuri [13] who takes into consideration the phase-lags of heat flux, temperature gradient and thermal displacement gradient. Chen and Gurtin [14], introduced the theory of two-temperatures (conductive temperature and thermodynamic temperature) heat conduction in the context of elastic bodies, then Youssef [15] extended this theory to generalized thermoelasticity. Fahmy [16] introduced three-temperature nonlinear generalized micropolar-magneto-thermoelasticity theory and developed a new boundary element technique for Modeling and Simulation of complex problems associated with this theory. Theory of micropolar elasticity [17, 18] has been developed for studying the mechanical behavior of polymers and elastomers and applied in many applications [19–24]. Then, Eringen [25] and Nowacki [26] extended it to micropolar thermoelasticity, and then implemented in various applications [27–29]. Because of strong nonlinearity of three-temperatures radiative heat conduction equations, the numerical solution and simulation of such problems are always difficult and require the development of new numerical schemes [30, 31]. In comparison with other numerical methods [32–34], the boundary element method has been successfully applied and performed for solving various applications [35–60]. The boundary element technique has been formulated in the context of micropolar thermoelasticity by Sladek and Sladek [61–63] and Huang and Liang [64]. Through the current paper, the term three-temperatures introduced for the first time in the field of nonlinear generalized micropolar thermoviscoelasticity. Recently, evolutionary algorithms [65, 66] have received much attention of researchers. The genetic algorithm (GA) can deal with the multi-objective and complex shapes problems. Also, it could reach an optimal solution with highly reduced computational cost.

The main aim of this article is to introduce a new theory called nonlinear generalized micropolar thermoviscoelasticity involving three temperatures. Because of strong nonlinearity, it is very difficult to solve the problems related to this theory analytically. Therefore, we propose a new boundary element model for simulation and optimization of three temperatures nonlinear generalized micropolar thermoviscoelastic problems associated with this theory. The genetic algorithm (GA) was implemented based on free form deformation (FFD) technique and nonuniform rational B-spline (NURBS) curve as an optimization technique for the considered problems. The numerical results demonstrate the validity and accuracy of our proposed model.

2. Formulation of the problem

The governing equations for three-temperature anisotropic generalized micropolar thermoviscoelasticity problems can be expressed as [58]

$$\sigma_{ij,j} + \rho F_i = \rho \ddot{u}_i \tag{1}$$

$$m_{ij,j} + \varepsilon_{ijk}\sigma_{jk} + \rho M_i = J\rho\ddot{\omega}_i \tag{2}$$

where

$$\sigma_{ij} = C_{ijkl} \aleph\, e\delta_{ij} + \check{\alpha}\left(u_{j,i} - \varepsilon_{ijk}\omega_k\right) - \beta_{ij}T_\alpha \left(C_{ijkl} = C_{klij} = C_{jikl}, \beta_{ij} = \beta_{ji}\right) \tag{3}$$

$$m_{ij} = \alpha\, \omega_{k,k}\delta_{ij} + \overline{\alpha}\omega_{i,j} + \overline{\overline{\alpha}}\omega_{j,i} \tag{4}$$

$$\epsilon_{ij} = \varepsilon_{ij} - \varepsilon_{ijk}(r_k - \omega_k), \varepsilon_{ij} = \frac{1}{2}\left(u_{i,j} + u_{j,i}\right), r_i = \frac{1}{2}\varepsilon_{ikl}u_{l,k} \tag{5}$$

The two-dimension three-temperature (2D-3T) radiative heat conduction equations can be expressed as [53]

$$c_e\frac{\partial T_e(r,\tau)}{\partial\tau} - \frac{1}{\rho}\nabla[\mathbb{K}_e\nabla T_e(r,\tau)] = -\mathbb{W}_{ei}(T_e - T_i) - \mathbb{W}_{ep}\left(T_e - T_p\right) \tag{6}$$

$$c_i\frac{\partial T_i(r,\tau)}{\partial\tau} - \frac{1}{\rho}\nabla[\mathbb{K}_i\nabla T_i(r,\tau)] = \mathbb{W}_{ei}(T_e - T_i) \tag{7}$$

$$\frac{4}{\rho}c_p T_p^3\frac{\partial T_p(r,\tau)}{\partial\tau} - \frac{1}{\rho}\nabla\left[\mathbb{K}_p\nabla T_p(r,\tau)\right] = \mathbb{W}_{ep}\left(T_e - T_p\right) \tag{8}$$

3. A new mathematical modelling of nonlinear generalized micropolar thermoviscoelasticity problem

With reference to a Cartesian coordinate system (x_1, x_2, x_3), we consider an anisotropic micropolar thermoviscoelastic structure occupies the region R which bounded by a closed surface S, and $S_i (i = 1,2,3,4)$ denotes subsets of S such that $S_1 + S_2 = S_3 + S_4 = S$.

3.1 BEM simulation for temperature field

The 2D-3T radiative heat conduction Eqs. (6)–(8) can be expressed as [53]

$$\nabla[\mathbb{K}_\alpha\nabla T_\alpha(r,\tau)] + \overline{\mathbb{W}}(r,\tau) = c_\alpha\rho\delta_1\frac{\partial T_\alpha(r,\tau)}{\partial\tau} \tag{9}$$

where

$$\overline{\mathbb{W}}(r,\tau) = \begin{cases} -\rho\,\mathbb{W}_{ei}(T_e - T_i) - \rho\,\mathbb{W}_{ep}\left(T_e - T_p\right), \alpha = e, \delta_1 = 1 \\ \rho\,\mathbb{W}_{ei}(T_e - T_i), \alpha = i, \delta_1 = 1 \\ \rho\,\mathbb{W}_{ep}\left(T_e - T_p\right), \alpha = p, \delta_1 = \frac{4}{\rho}T_p^3 \end{cases} \tag{10}$$

and

$$\mathbb{W}_{ei} = \rho\mathbb{A}_{ei}T_e^{-2/3}, \mathbb{W}_{ep} = \rho\mathbb{A}_{ep}T_e^{-1/2}, \mathbb{K}_\alpha = \mathbb{A}_\alpha T_\alpha^{5/2}, \alpha = e, i, \mathbb{K}_p = \mathbb{A}_p T_p^{3+\mathbb{B}} \tag{11}$$

where parameters c_α, $\mathbb{A}_\alpha(\alpha = e, i, p)$, $\mathbb{B}$, $\mathbb{A}_{ei}$, $\mathbb{A}_{ep}$ are constant inside each subdomain, but they are discontinuous on the interfaces between subdomains.

The total energy of unit mass can be described by

$$P = P_e + P_i + P_p, P_e = c_e T_e, P_i = c_i T_i, P_p = \frac{1}{\rho} c_p T_p^4 \quad (12)$$

Initial and boundary conditions can be written as

$$T_\alpha(x, y, 0) = T_\alpha^0(x, y) = g_1(x, \tau) \quad (13)$$

$$\mathbb{K}_\alpha \frac{\partial T_\alpha}{\partial n}\bigg|_{\Gamma_1} = 0, \alpha = e, i, T_p|_{\Gamma_1} = g_2(x, \tau) \quad (14)$$

$$\mathbb{K}_\alpha \frac{\partial T_\alpha}{\partial n}\bigg|_{\Gamma_2} = 0, \alpha = e, i, p \quad (15)$$

we use the time-dependent fundamental solution which is a solution of the following differential equation

$$D\nabla^2 T_\alpha + \frac{\partial T_\alpha^*}{\partial n} = -\delta(r - p_i)\delta(\tau - r), D = \frac{\mathbb{K}_\alpha}{\rho c} \quad (16)$$

In which the points p_i are the singularities, where the temperatures are not defined there. Singular integrals are those whose kernels are not defined at the singularities on the integration domain R. They are defined by eliminating a small space including the singularity, and obtaining the limit when this small space tends to zero [40, 46].

The boundary integral equation corresponding to our considered heat conduction can be written as in Fahmy [46–48] as follows

$$CT_\alpha = \frac{D}{\mathbb{K}_\alpha} \int_O^\tau \int_S \left[T_\alpha q^* - T_\alpha^* q\right] dS\, d\tau + \frac{D}{\mathbb{K}_\alpha} \int_O^\tau \int_R bT_\alpha^*\, dR\, d\tau + \int_R T_\alpha^i T_\alpha^* |_{\tau=0} \quad (17)$$

which can be expressed in the following form [53].

$$CT_\alpha = \int_S \left[T_\alpha q^* - T_\alpha^* q\right] dS - \int_R \frac{\mathbb{K}_\alpha}{D} \frac{\partial T_\alpha^*}{\partial \tau} T_\alpha\, dR \quad (18)$$

The time derivative of temperature T_α can be approximated as

$$\frac{\partial T_\alpha}{\partial \tau} \cong \sum_{j=1}^{N} f^j(r)^j a^j(\tau). \quad (19)$$

where $f^j(r)$ and $a^j(\tau)$ are known functions and unknown coefficients, respectively.

Also, we assume that $\widehat{T}_\alpha^j$ is a solution of

$$\nabla^2 \widehat{T}_\alpha^j = f^j \quad (20)$$

Thus, Eq. (18) results in the following boundary integral equation [53]

$$C\,T_\alpha = \int_S \left[T_\alpha q^* - T_\alpha^* q\right] dS + \sum_{j=1}^{N} a^j(\tau) D^{-1} \left(C\widehat{T}_\alpha^j - \int_S \left[T_\alpha^j q^* - \widehat{q}^j T_\alpha^*\right] dS \right) \quad (21)$$

where

$$\widehat{q}^{j} = -\mathbb{K}_{\alpha}\frac{\partial \widehat{T}_{\alpha}^{j}}{\partial n} \tag{22}$$

and

$$a^{j}(\tau) = \sum_{i=1}^{N} f_{ji}^{-1}\frac{\partial T(r_i,\tau)}{\partial \tau} \tag{23}$$

In which the entries of f_{ji}^{-1} are the coefficients of F^{-1} which described in [34].

$$\{F\}_{ji} = f^{j}(r_i) \tag{24}$$

The boundary integral discretization scheme has been applied to (21) with the use of (23), we get [53]

$$C\,\dot{T}_{\alpha} + H\,T_{\alpha} = G\,Q \tag{25}$$

where T_{α} and Q are temperature, heat flux vectors and internal heat generation vectors, respectively.

The diffusion matrix can be defined as

$$C = -\left[H\,\widehat{T}_{\alpha} - G\,\widehat{Q}\right]F^{-1}D^{-1} \tag{26}$$

with

$$\left\{\widehat{T}\right\}_{ij} = \widehat{T}^{j}(x_i) \tag{27}$$

$$\left\{\widehat{Q}\right\}_{ij} = \widehat{q}^{j}(x_i) \tag{28}$$

For solving (25) numerically, we interpolate the functions T_{α} and q as

$$T_{\alpha} = (1-\theta)T_{\alpha}^{m} + \theta\,T_{\alpha}^{m+1} \tag{29}$$

$$q = (1-\theta)q^{m} + \theta\,q^{m+1} \tag{30}$$

where $0 \le \theta = \frac{\tau-\tau^{m}}{\tau^{m+1}-\tau^{m}} \le 1$.

The time derivative of (29) can be expressed as

$$\dot{T}_{\alpha} = \frac{dT_{\alpha}}{d\theta}\frac{d\theta}{d\tau} = \frac{T_{\alpha}^{m+1} - T_{\alpha}^{m}}{\tau^{m+1} - \tau^{m}} = \frac{T_{\alpha}^{m+1} - T_{\alpha}^{m}}{\Delta\tau^{m}} \tag{31}$$

By substituting from Eqs. (29)–(31) into Eq. (25), we obtain

$$\left(\frac{C}{\Delta\tau^{m}} + \theta H\right)T_{\alpha}^{m+1} - \theta G Q^{m+1} = \left(\frac{C}{\Delta\tau^{m}} - (1-\theta)H\right)T_{\alpha}^{m} + (1-\theta)GQ^{m} \tag{32}$$

Making use of initial conditions and boundary conditions at $\Delta\tau^{m}$ and considering the previous time step solution as initial values for next step, we get

$$\mathbb{a}X = \mathbb{b} \tag{33}$$

The Adaptive Smoothing and Prolongation Algebraic Multigrid (aSP-AMG) method, which uses an adaptive Factorized Sparse Approximate Inverse (aFSAI) [67] preconditioner as high performance technique that has been implemented efficiently in Matlab (R2018a) for solving the resulting simultaneous linear algebraic systems (33).

3.2 BEM simulation for micropolar thermoviscoelastic fields

According to the weighted residual method, we can write the differential Eqs. (1) and (2) in the following integral form

$$\int_R (\sigma_{ij,j} + U_i) u_i^* \, dR = 0 \tag{34}$$

$$\int_R (m_{ij,j} + \varepsilon_{ijk}\sigma_{jk} + V_i)\omega_i^* \, dR = 0 \tag{35}$$

where

$$U_i = \rho F_i - \rho \ddot{u}_i, V_i = \rho(M_i - J\ddot{\omega}_i) \tag{36}$$

The boundary conditions are

$$u_i = \overline{u}_i \text{ on } S_1 \tag{37}$$

$$\lambda_i = \sigma_{ij} n_j = \overline{\lambda}_i \text{ on } S_2 \tag{38}$$

$$\omega_i = \overline{\omega}_i \text{ on } S_3 \tag{39}$$

$$\mu_i = m_{ij} n_j = \overline{\mu}_i \text{ on } S_4 \tag{40}$$

By integrating by parts the first term of Eqs. (34) and (35), we obtain

$$-\int_R \sigma_{ij} u_{i,j}^* \, dR + \int_R U_i u_i^* \, dR = -\int_{S_2} \lambda_i u_i^* \, dS \tag{41}$$

$$-\int_R m_{ij}\omega_{i,j}^* \, dR + \int_R \varepsilon_{ijk}\sigma_{jk}\omega_i^* \, dR + \int_R V_i \omega_i^* \, dR = -\int_{S_4} \mu_i \omega_i^* \, dS \tag{42}$$

On the basis of Huang and Liang [64], we can write

$$\begin{aligned} -\int_R \sigma_{ij,j} u_i^* \, dR + \int_R (m_{ij,j} + \varepsilon_{ijk}\sigma_{jk})\omega_i^* \, dR + \int_R U_i u_i^* \, dR + \int_R V_i \omega_i^* \, dR \\ = \int_{S_2} (\lambda_i - \overline{\lambda}_i) u_i^* \, dS + \int_{S_1} (\overline{u}_i - u_i)\lambda_i^* \, dS + \int_{S_4} (\mu_i - \overline{\mu}_i)\omega_i^* \, dS + \int_{S_3} (\overline{\omega}_i - \omega_i)\mu_i^* \, dS \end{aligned} \tag{43}$$

By integrating by parts, the left-hand side of (43) can be written as

$$\begin{aligned} -\int_R \sigma_{ij}\varepsilon_{ij}^* \, dR - \int_R m_{ij,j}\omega_{i,j}^* \, dR + \int_R U_i u_i^* \, dR + \int_R V_i \omega_i^* \, dR \\ = -\int_{S_2} \overline{\lambda}_i \, u_i^* \, dS - \int_{S_1} \lambda_i u_i^* \, dS + \int_{S_1} (\overline{u}_i - u_i)\lambda_i^* \, dS - \int_{S_4} \overline{\mu}_i \omega_i^* \, dS - \int_{S_3} \mu_i \omega_i^* \, dS \\ + \int_{S_3} (\overline{\omega}_i - \omega_i)\mu_i^* \, dS \end{aligned} \tag{44}$$

According to Eringen [68], the elastic and couple stresses can be written in the following form

$$\sigma_{ij} = \mathbb{A}_{ijkl}\varepsilon_{kl}, m_{ij} = \mathbb{B}_{ijkl}\omega_{k,l} \tag{45}$$

where $\mathbb{A}_{ijkl} = \mathbb{A}_{klij}$ and $\mathbb{B}_{ijkl} = \mathbb{B}_{klij}$ as shown in [68].

Hence, Eq. (44) can be re-expressed as [53]

$$\begin{aligned}
&-\int_R \sigma_{ij}^* \varepsilon_{ij}\, dR - \int_R m_{ij,j}^* \omega_{i,j}\, dR + \int_R U_i u_i^*\, dR + \int_R V_i \omega_i^*\, dR \\
&= -\int_{S_2} \overline{\lambda}_i\, u_i^*\, dS - \int_{S_1} \lambda_i u_i^*\, dS + \int_{S_1} (\overline{u}_i - u_i)\lambda_i^*\, dS - \int_{S_4} \overline{\mu}_i \omega_i^*\, dS - \int_{S_3} \mu_i \omega_i^*\, dS \\
&\quad + \int_{S_3} (\overline{\omega}_i - \omega_i)\mu_i^*\, dS
\end{aligned} \tag{46}$$

By applying integration by parts again, the left-hand side of (46) can be written as [53]

$$\begin{aligned}
\int_R \sigma_{ij,j}^* u_i\, dR + \int_R \left(m_{ij,j}^* + \varepsilon_{ijk}\sigma_{jk}^*\right)\omega_i\, dR = -\int_S u_i^* \lambda_i\, dS - \int_S \omega_i^* \mu_i\, dS + \int_S \lambda_i^* u_i\, dS \\
+ \int_S \mu_i^* \omega_i\, dS
\end{aligned} \tag{47}$$

The obtained weighting functions for $U_i = \Delta^n$ and $V_i = 0$ along e_l were first used as follows:

$$\sigma_{lj,j}^* + \Delta^n e_l = 0 \tag{48}$$

$$m_{ij,j}^* + \varepsilon_{ijk}\sigma_{jk}^* = 0 \tag{49}$$

According to Dragos [69], the fundamental solutions can be written as

$$u_i^* = u_{li}^* e_l, \omega_i^* = \omega_{li}^* e_l, \lambda_i^* = \lambda_{li}^* e_l, \mu_i^* = \mu_{li}^* e_l, \tag{50}$$

The obtained weighting functions for $U_i = 0$ and $V_i = \Delta^n$ along e_l were next used as follows:

$$\sigma_{ij,j}^{**} = 0 \tag{51}$$

$$m_{lj,j}^{**} + \varepsilon_{ljk}\sigma_{jk}^{**} + \Delta^n e_l = 0 \tag{52}$$

The fundamental solutions that have been obtained analytically by Dragos [69] can be written as

$$u_i^* = u_{li}^{**} e_l, \omega_i^* = \omega_{li}^{**} e_l, \lambda_i^* = \lambda_{li}^{**} e_l, \mu_i^* = \mu_{li}^{**} e_l, \tag{53}$$

Using the above two sets of weighting functions into (47) we have

$$C_{li}^n u_i^n = -\int_S \lambda_{li}^* u_i dS - \int_S \mu_{li}^* \omega_i dS + \int_S u_{li}^* \lambda_i\, dS + \int_S \omega_{li}^* \mu_i\, dS \tag{54}$$

$$C_{li}^n \omega_i^n = -\int_S \lambda_{li}^{**} u_i dS - \int_S \mu_{li}^{**} \omega_i dS + \int_S u_{li}^{**} \lambda_i\, dS + \int_S \omega_{li}^{**} \mu_i\, dS \tag{55}$$

Thus, we can write

$$\mathbb{C}^n\mathbb{q}^n = -\int_S \mathbb{p}^* \, \mathbb{q}dS + \int_S \mathbb{q}^* \, \mathbb{p}dS \tag{56}$$

where

$$C^n = \begin{bmatrix} C_{11} & C_{12} & 0 \\ C_{21} & C_{22} & 0 \\ 0 & 0 & 0 \end{bmatrix}, \mathbb{q}^* = \begin{bmatrix} u_{11}^* & u_{12}^* & \omega_{13}^* \\ u_{21}^* & u_{22}^* & \omega_{23}^* \\ u_{31}^{**} & u_{32}^{**} & \omega_{33}^{**} \end{bmatrix}, \mathbb{p}^* = \begin{bmatrix} \lambda_{11}^* & \lambda_{12}^* & \mu_{13}^* \\ \lambda_{21}^* & \lambda_{22}^* & \mu_{23}^* \\ \lambda_{31}^{**} & \lambda_{32}^{**} & \mu_{33}^{**} \end{bmatrix},$$

$$\mathbb{q} = \begin{bmatrix} u_1 \\ u_2 \\ \omega_3 \end{bmatrix}, \mathbb{p} = \begin{bmatrix} \lambda_1 \\ \lambda_2 \\ \mu_3 \end{bmatrix}$$

In order to solve (56) numerically, we construct the following functions

$$\mathrm{q} = \psi\, \mathrm{q}^j, \mathrm{p} = \psi\, \mathrm{p}^j \tag{57}$$

substituting above functions into (56) and discretizing the boundary, we obtain

$$\mathbb{C}^n\mathbb{q}^n = \sum_{j=1}^{N_e}\left[-\int_{\Gamma_j} \mathbb{p}^* \psi \, d\Gamma\right]\mathbb{q}^j + \sum_{j=1}^{N_e}\left[\int_{\Gamma_j} \mathbb{q}^* \psi \, d\Gamma\right]\mathbb{p}^j \tag{58}$$

Equation after integration may be expressed as

$$\mathbb{C}^i\mathbb{q}^i = -\sum_{j=1}^{N_e} \widehat{\mathbb{H}}^{ij}\mathbb{q}^j + \sum_{j=1}^{N_e} \widehat{\mathbb{G}}^{ij}\mathbb{p}^j \tag{59}$$

which can be expressed as

$$\sum_{j=1}^{N_e} \mathbb{H}^{ij}\mathbb{q}^j = \sum_{j=1}^{N_e} \widehat{\mathbb{G}}^{ij}\mathbb{p}^j \tag{60}$$

where

$$\mathbb{H}^{ij} = \begin{cases} \widehat{\mathbb{H}}^{ij} \text{ if } i \neq j \\ \widehat{\mathbb{H}}^{ij} + \mathbb{C}^i \text{ if } i = j \end{cases} \tag{61}$$

Thus, we can write the following system of matrix equation as

$$\mathbb{H}\mathbb{Q} = \mathbb{G}\mathbb{P} \tag{62}$$

Hence, we get the following system of linear algebraic equations

$$\mathbb{A}\,\mathbb{X} = \mathbb{B} \tag{63}$$

4. A new boundary element technique for simulation and optimization of solid deformable bodies under different loads

In order to solve (63), we apply adaptive smoothing and prolongation algebraic multigrid (aSP-AMG) based on adaptive Factorized Sparse Approximate Inverse (aFSAI) as described in [67] for solving the resulting simultaneous linear algebraic system (63) in Matlab (R2018a).

B-spline basis functions are used as weights in the same manner as Bézier basis functions. Spline curves can be expressed in terms of $k-$ order B-spline basis function. All B-spline basis functions are assumed to have their domain on [0,1]. B-splines basis functions are a more general type of curve than Bezier curves, where each control point P_i of $i+1$ control points $(P_0, P_1, P_2, \dots, P_i)$ is connected with a basis function $N_{i,k}$, the knots are the points that subdivide the domain [0,1] into knot spans. Also, each B-spline basis function is non-zero on the entire interval.

The efficiency of our numerical modeling technique has been improved using a nonuniform rational B-spline curve (NURBS) to decrease the computation time and model optimized boundary where it reduces the number of control points and provides the flexibility to design a large variety of shapes.

The considered NURBS can be defined as follows

$$C(t) = \frac{\sum_{i=0}^{n} N_{i,o}(t)\varpi_i P_i}{\sum_{i=0}^{n} N_{i,o}(t)\varpi_i} \tag{64}$$

where $N_{i,o}(t)$ and ϖ_i are the B-spline basis functions of order o and the weights of control points P_i, respectively.

The genetic algorithm greatly reduces computing time and computer memory of achieving an optimum solution, so, it can be used for solving multi-objective problems without needing to calculate the sensitivities. The profiles of the considered objects are represented based on the free form deformation (FFD) technique, where the FFD control points are considered as the genes and then the profiles of chromosomes are defined by the sequence of genes. The population is constructed by many individuals (chromosomes), where the fitness functions are evaluated by using the BEM.

Two criteria can be implemented during shape optimization of the solid bodies [70]

I. The minimum global compliance:

$$\mathcal{F} = \frac{1}{2}\int_S (\lambda \cdot u)\, dS \tag{65}$$

II. The minimum boundary equivalent stresses

$$\mathcal{F} = \int_S \left(\frac{\sigma_{ij}}{\sigma_0}\right)^n dS \tag{66}$$

σ_{ij}, σ_0 and n are equivalent boundary stresses, reference stress and natural number, respectively, where the greater value of n increases the speed of convergence of the functional (66). By minimizing the functional (66) σ_{ij} are closer to σ_0.

In order to find the optimal boundary conditions for temperature the following functional can be applied

$$\mathcal{F} = \int_S \left(\frac{u}{u_0}\right)^n dS \tag{67}$$

where u and u_0 are boundary displacement and reference displacement, respectively. Minimization of the functional (67) reduces displacements on the selected part of the boundary.

In order to identify unknown inner boundary, we use the following functional

$$\mathcal{F} = \delta \sum_{k=1}^{M} \left(u^k - \widehat{u}^k \right) + \eta \sum_{l=1}^{N} \left(T^l - \widehat{T}^l \right) \tag{68}$$

where $\widehat{u}^k$ and $\widehat{T}^l$ are measured displacements and temperatures in boundary points k and l respectively, u^k and T^l are computed displacements and temperatures in boundary points k and l respectively, δ and η are weight coefficients, and M and N are numbers of sensors.

5. Numerical examples and discussion of results

For illustration of the theoretical results of our proposed technique from the preceding sections, two numerical examples are analyzed below. The first example is the cantilever beam with inferior corner load, the second is the Michell-type structure, where the material has the following physical data [58]:

The elasticity tensor

$$C_{ijkl} = \begin{pmatrix} 60.23 & 18.67 & 18.96 & -7.69 & 15.60 & -25.28 \\ 18.67 & 21.26 & 9.36 & -3.74 & 4.21 & -8.47 \\ 18.96 & 9.36 & 47.04 & -8.82 & 15.28 & -8.31 \\ -7.69 & -3.74 & -8.82 & 10.18 & -9.54 & 5.69 \\ 15.60 & 4.21 & 15.28 & -9.54 & 21.19 & -8.54 \\ -25.28 & -8.47 & -8.31 & 5.69 & -8.54 & 20.75 \end{pmatrix} \tag{69}$$

$p = 25$ MPa, and $\Delta t = 0.0006$ s.

Example 1. Cantilever beam structure.

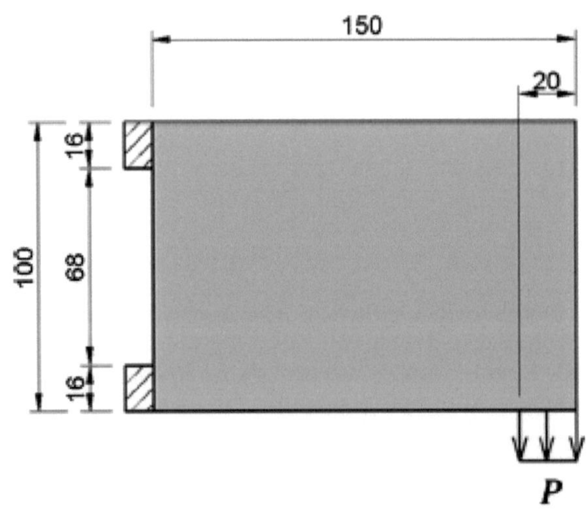

Figure 1.
Cantilever beam structure geometry.

As a practical example, the proposed algorithm is applied on the cantilever beam with inferior corner load P = 100 N/mm. The geometry of the cantilever beam is illustrated in **Figure 1**. The initial boundary mesh of the cantilever beam composed of 14 quadratic elements is also illustrated in **Figure 2**. The BEM grid is composed of 76 nodes along x direction and 51 nodes along y direction. These mesh parameters were obtained after convergence analysis. In the process of optimization, the cantilever beam structure optimization results are presented in **Figure 3** from initial to final structure for different iterations.

The present measured boundary element method (BEM) optimization results of the first example are compared in **Figure 4** with measured finite difference method (FDM) optimization results obtained by Itzá et al. [71] and measured finite element method (FEM) optimization results obtained using the software package COMSOL Multiphysics, version 5.4. It is clear from this figure that the BEM results obtained by the proposed technique are in excellent agreement with the FDM results [71] and FEM results of the COMSOL Multiphysics.

Table 1 shows that our proposed BEM modeling of cantilever beam with inferior corner load drastically reduces the manpower needed for modeling and computer resources needed for the calculation in comparison with the calculated results based on the FDM and FEM.

Example 2. Michell-type structure.

As application example, we use a beam with a mid-span load (P = 100 N/mm) (Michell-type structure) as shown in **Figure 5**. The initial boundary mesh of the Michell-type structure composed of 40 quadratic discontinuous elements is also

Figure 2.
Initial boundary of the cantilever beam structure.

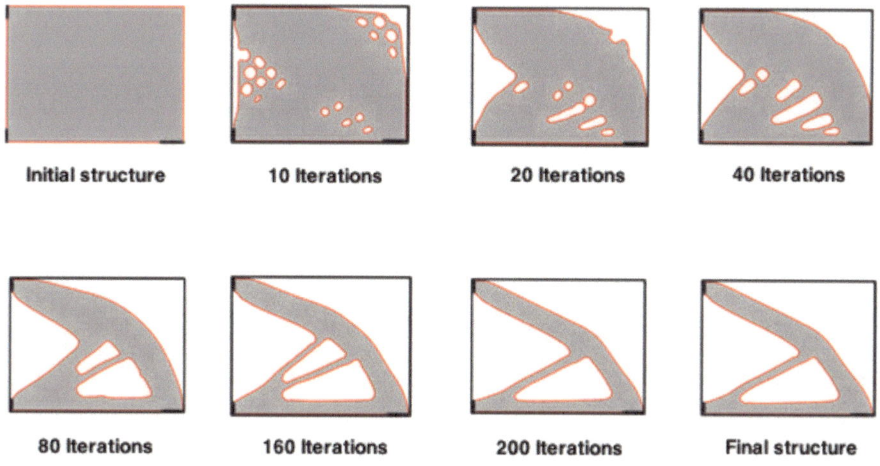

Figure 3.
Cantilever beam optimization process from initial to final structure for different iterations.

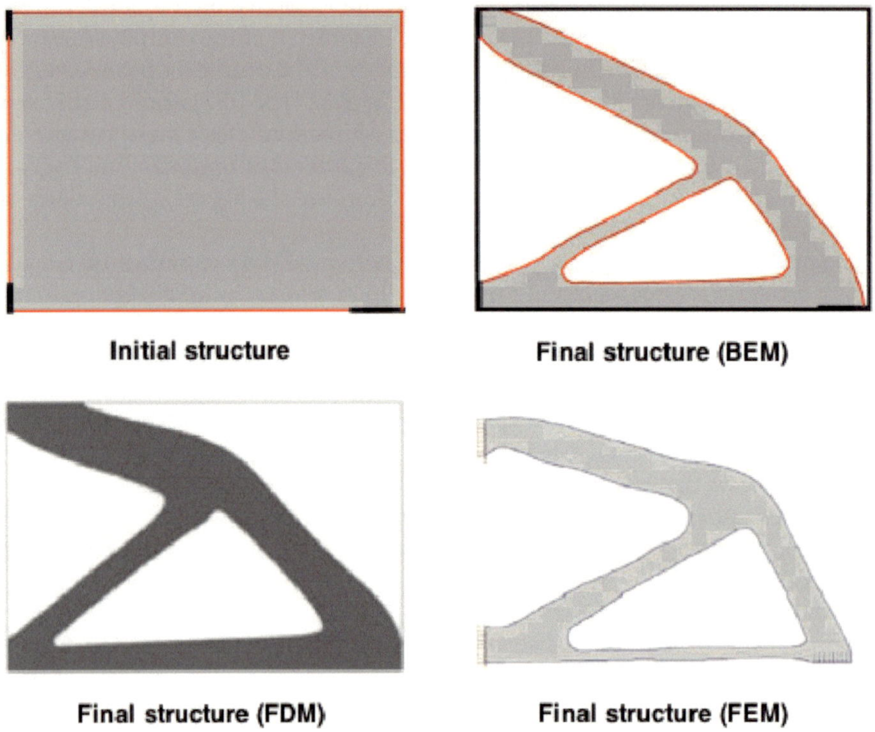

Figure 4.
Final cantilever beam structure for BEM, FDM and FEM.

	FDM	FEM	BEM
Number of Nodes	24000	27000	77
Number of Elements	8850	9650	40
CPU-Time [min.]	120	130	2
Memory [MByte]	75	85	0.4
Disc Space [MByte]	130	150	0
Accuracy of Results [%]	2.1	2.4	1.4

Table 1.
Comparison of computer resources needed for FDM, FEM and BEM modeling of cantilever beam structure.

illustrated in **Figure 6**. The BEM grid is composed of 76 nodes along x direction, and 51 nodes along y direction. This grid density was obtained after convergence analysis.

Figure 7 shows the cantilever beam optimization process from initial to final structure for different iterations.

The optimization results of the second example obtained with the proposed BEM are compared in **Figure 8** with FDM optimization results [71] and FEM optimization results of COMSOL Multiphysics software, version 5.4. It is clear from this figure that our BEM results obtained by the proposed technique are in excellent agreement with the FDM and FEM results.

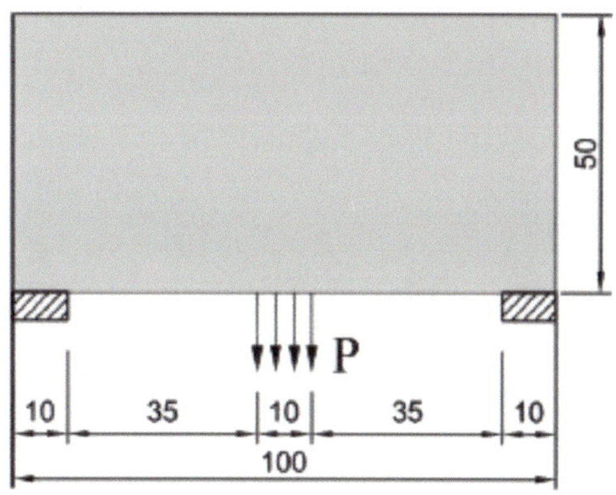

Figure 5.
Michel-type structure geometry.

Figure 6.
Initial boundary of the Michel-type structure.

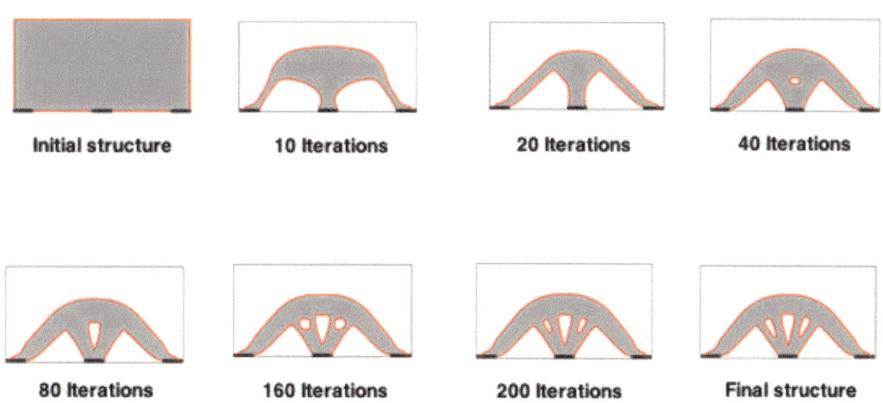

Figure 7.
Michell-type structure optimization process from initial to final structure for different iterations.

Table 2 shows that our proposed BEM modeling of Michell-type structure dramatically reduces the computer resources necessary to calculate our proposed modeling in comparison with the calculated results based on the FDM and FEM.

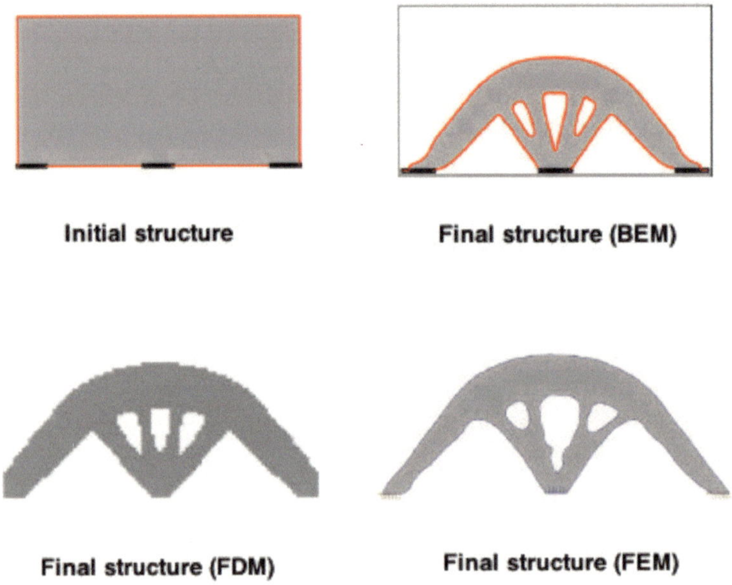

Figure 8.
Final Michell-type structure for BEM, FDM and FEM.

	FDM	FEM	BEM
Number of Nodes	26000	28000	30
Number of Elements	8950	9850	30
CPU-Time [min.]	130	140	2
Memory [MByte]	80	90	0.5
Disc Space [MByte]	160	180	0
Accuracy of Results [%]	2.2	2.5	1.5

Table 2.
Comparison of computer resources needed for FDM, FEM and BEM modeling of Michell-type structure.

6. Conclusion

In the present paper, we propose a new theory called nonlinear micropolar thermoviscoelasticity involving three temperatures. A new mathematical modeling of nonlinear generalized micropolar thermoviscoelasticity problem. A new boundary element technique for simulation and optimization problems of mechanics of solid deformable bodies is implemented based on genetic algorithm (GA), free form deformation (FFD) method and nonuniform rational B-spline curve (NURBS) as the global optimization technique for solving complex simulation and optimization problems associated with the proposed theory. FFD is an efficient and versatile parameterization technique for treating shape optimization problems with complex shapes. It is implemented for simulation and optimization of the shape. In the formulation of the considered problem, solutions are obtained for specific arbitrary parameters which are the control points positions in the considered problem, the profiles of the considered objects are determined by FFD method, where the FFD control points positions are treated as genes, and then the chromosomes profiles are defined with the genes sequence. The population is founded by a number of individuals (chromosomes), where the objective functions of individuals are determined by the boundary element method (BEM). Due to the large amount of computer resources required by the FDM and FEM, our proposed BEM model can be applied to a wide range of simulation and optimization problems related with our proposed theory. The numerical results demonstrate the validity, accuracy and efficiency of our proposed technique.

Nomenclature

$\alpha, \overline{\alpha}, \overline{\overline{\alpha}}, \breve{\alpha}$	micro-polar constants
β_{ij}	stress–temperature coefficients
δ_{ij}	Kronecker delta $(i, j = 1, 2)$
δ, η	weight coefficients
ε_{ij}	strain tensor
ε_{ijk}	alternate tensor
ϵ_{ij}	micro-strain tensor
λ	tractions
ϑ_0	viscoelastic relaxation time
ϖ	weights of control points
ρ	material density
σ_{ij}	force stress tensor
σ_0	reference stress
τ	time
τ_0, τ_1, τ_2	relaxation times
ω_i	micro-rotation vector
$\aleph$	$= \left(1 + \vartheta_0 \frac{\partial}{\partial \tau}\right)$ viscoelastic constant
b	internal heat generation vector
c	specific heat capacity
C_{ijkl}	constant elastic moduli
e	$= \varepsilon_{kk} = \epsilon_{kk}$ dilatation
$\mathfrak{e}_{lij}$	piezoelectric tensor
F_i	mass force vector
J	micro-inertia coefficient
$\mathbb{J}$	current density vector

$\mathbb{K}_e$	electron conductive coefficients
$\mathbb{K}_i$	ion conductive coefficients
$\mathbb{K}_r$	phonon conductive coefficients
k, l	boundary points
M, N	sensors numbers
M_i	mass couple vector
m_{ij}	couple stress tensor
P	total energy of unit mass
P_e	$= c_e T_e$ Electron energy
P_i	$= c_i T_i$ Ion energy
P_r	$= \frac{1}{\rho} c_r T_r^4$ Phonon energy
$\mathbb{P}$	values vector of tractions and couple stress
$\mathrm{p_i}$	singular points
Q	heat flux vectors
$\mathbb{Q}$	values vector of displacements and microrotations
R	problem's boundary
S	problem's domain
T_e	electron temperature
T_i	ion temperature
T_r	phonon temperature
u	boundary displacement
u_i	displacement vector
u_0	reference displacement
u^k	computed displacements
$\hat{u}^k$	measured displacements
$\mathbb{W}_{\mathrm{ei}}$	electron-ion energy coefficient
$\mathbb{W}_{\mathrm{er}}$	electron–phonon energy coefficient

Author details

Mohamed Abdelsabour Fahmy
Faculty of Computers and Informatics, Suez Canal University, Ismailia, Egypt

*Address all correspondence to: mohamed_fahmy@ci.suez.edu.eg

References

[1] Duhamel J. Some memoire sur les phenomenes thermo-mechanique. Journal de l'École Polytechnique. 1837; **15**:1-57

[2] Neumann F. Vorlesungen Uber die theorie der elasticitat. Brestau: Meyer; 1885

[3] Biot M. Thermoelasticity and irreversible thermo-dynamics. Journal of Applied Physics. 1956;**27**:249-253

[4] Lord HW, Shulman Y. A generalized dynamical theory of thermoelasticity. Journal of the Mechanics and Physics of Solids. 1967;**15**:299-309

[5] Green AE, Lindsay KA. Thermoelasticity. Journal of Elasticity. 1972;**2**:249-253

[6] Green AE, Naghdi PM. On undamped heat waves in an elastic solid. Journal of Thermal Stresses. 1992;**15**: 253-264

[7] Green AE, Naghdi PM. Thermoelasticity without energy dissipation. Journal of Elasticity. 1993; **31**:189-208

[8] Hetnarski RB, Ignaczak J. Soliton-like waves in a low-temperature nonlinear thermoelastic solid. International Journal of Engineering Science. 1996;**34**: 1767-1787

[9] Tzou DY. A unified approach for heat conduction from macro to micro scales. ASME Journal of Heat Transfer. 1995;**117**:8-16

[10] Tzou DY. The generalized lagging response in small-scale and high-rate heating. International Journal of Heat and Mass Transfer. 1995;**38**:3231-3240

[11] Tzou DY. Macro-to Microscale Heat Transfer: The Lagging Behavior. New York: Taylor & Francis; 1997

[12] Chandrasekharaiah DS. Hyperbolic thermoelasticity: A review of recent literature. Applied Mechanics Reviews. 1998;**51**:705-729

[13] Choudhuri SKR. On a thermoelastic three-phase-lag model. Journal of Thermal Stresses. 2007;**30**:231-238

[14] Chen PJ, Gurtin ME. On a theory of heat conduction involving two temperatures. Zeitschrift für Angewandte Mathematik und Physik. 1968;**19**:614-627

[15] Youssef H. Theory of two-temperature generalized thermoelasticity. IMA Journal of Applied Mathematics. 2006;**71**:383-390

[16] Fahmy MA. A new boundary element strategy for modeling and simulation of three-temperature nonlinear generalized micropolar-magneto-thermoelastic wave propagation problems in FGA structures. Engineering Analysis with Boundary Elements. 2019;**108**:192-200

[17] Eringen AC, Suhubi ES. Nonlinear theory of simple micro-elastic solid-I. International Journal of Engineering Science. 1964;**2**:189-203

[18] Eringen AC. Linear theory of micropolar elasticity. Journal of Mathematics and Mechanics. 1966;**15**: 909-923

[19] Tauchert TR Jr, Claus WD, Ariman T. The linear theory of micropolar thermoelasticity. International Journal of Engineering Science. 1968;**6**:36-47

[20] Eringen AC. Foundation of Micropolar Thermoelasticity, Courses and Lectures No. 23. Wein, Vienna: CISM International Centre for Mechanical Sciences/Springer-Verlag; 1970

[21] Tauchert TR. Thermal stresses in micropolar elastic solids. Acta Mech. 1971;**11**:155-169

[22] Dhaliwal RS. The steady-state axisymmetric problem of micropolar thermoelasticity. Archives of Mechanics. 1971;**23**:705-714

[23] Nowacki W, Olszak W. Micropolar Thermoelasticity. Wien, Vienna: CISM International Centre for Mechanical Sciences/Springer-Verlag; 1974

[24] Hetnarski RB, Eslami MR. Thermal Stresses—Advanced Theory and Applications. Netherlands: Springer-Verlag; 2008

[25] Eringen AC. Foundations of Micropolar Thermoelasticity. International Centre for Mechanical Sciences Book Series. Wien GMBH: Springer-Verlag; 1970

[26] Nowacki W. Couple-stresses in the theory of thermoelasticity. In: Proceedings IUTAM Symposia on Irreversible Aspects of Continuum Mechanics and Transfer of Physical Characteristics in Moving Fluids. Springer; 1968. pp. 259-278

[27] Marin M. On existence and uniqueness in thermoelasticity of micropolar bodies. Comptes Rendus de l'Académie des Sciences - Series I - Mathematics;**1995321**:475-480

[28] Ciarletta M. A theory of micropolar thermoelasticity without energy dissipation. Journal of Thermal Stresses. 1999;**22**:581-594

[29] Chirila A, Agarwal RP, Marin M. Proving uniqueness for the solution of the problem of homogeneous and anisotropic micropolar thermoelasticity. Boundary Value Problems. 2017;**2017**:3

[30] Nie C, Yu H. A Raviart–Thomas mixed finite element scheme for the two-dimensional three-temperature heat conduction problems. International Journal for Numerical Methods in Engineering. 2017;**111**:983-1000

[31] Hu Q, Zhao L. Domain decomposition preconditioners for the system generated by discontinuous Galerkin discretization of 2D-3T heat conduction equations. Communications in Computational Physics. 2017;**22**: 1069-1100

[32] Yi CZ, Zhang XW, Yan HX, Jin B. Finite element simulation and the application of amphoteric pH-sensitive hydrogel. International Journal of Applied Mechanics. 2017;**9**:1750063

[33] Eskandari AH, Baghani M, Sohrabpour S. A time-dependent finite element formulation for thick shape memory polymer beams considering shear effects. International Journal of Applied Mechanics. 2018;**10**:1850043

[34] Taghizadeh DM, Darijani H. Mechanical behavior modeling of hyperelastic transversely isotropic materials based on a new polyconvex strain energy function. International Journal of Applied Mechanics. 2018;**10**: 1850104

[35] Haghighat AE, Binesh SM. Domain decomposition algorithm for coupling of finite element and boundary element methods. Arabian Journal for Science and Engineering. 2014;**39**:3489-3497

[36] Fahmy MA. Boundary element algorithm for modeling and simulation of dual-phase lag bioheat transfer and biomechanics of anisotropic soft tissues. International Journal of Applied Mechanics. 2018;**10**:1850108

[37] Fahmy MA. Thermoelastic stresses in a rotating non- homogeneous anisotropic body. Numerical Heat Transfer, Part A: Applications. 2008;**53**: 1001-1011

[38] Fahmy MA, El-Shahat TM. The effect of initial stress and inhomogeneity on the thermoelastic stresses in a rotating anisotropic solid. Archive of Applied Mechanics. 2008;**78**: 431-442

[39] Fahmy MA. A time-stepping DRBEM for magneto-thermo-viscoelastic interactions in a rotating nonhomogeneous anisotropic solid. International Journal of Applied Mechanics. 2011;**3**:1-24

[40] Fahmy MA. A time-stepping DRBEM for the transient magneto-thermo-visco-elastic stresses in a rotating non-homogeneous anisotropic solid. Engineering Analysis with Boundary Elements. 2012;**36**:335-345

[41] Fahmy MA. Transient magneto-thermoviscoelastic plane waves in a non-homogeneous anisotropic thick strip subjected to a moving heat source. Applied Mathematical Modelling. 2012; **36**:4565-4578

[42] Fahmy MA. Numerical modeling of transient magneto-thermo-viscoelastic waves in a rotating nonhomogeneous anisotropic solid under initial stress. International Journal of Modeling, Simulation and Scientific Computing. 2012;**3**:1250002

[43] Fahmy MA. The effect of rotation and inhomogeneity on the transient magneto-thermoviscoelastic stresses in an anisotropic solid. ASME Journal of Applied Mechanics. 2012;**79**:1015

[44] Fahmy MA. Transient magneto-thermo-viscoelastic stresses in a rotating nonhomogeneous anisotropic solid with and without a moving heat source. Journal of Engineering Physics and Thermophysics. 2012;**85**:950-958

[45] Fahmy MA. Transient magneto-thermo-elastic stresses in an anisotropic viscoelastic solid with and without moving heat source. Numerical Heat Transfer, Part A: Applications. 2012;**61**: 547-564

[46] Fahmy MA. Implicit-explicit time integration DRBEM for generalized magneto-thermoelasticity problems of rotating anisotropic viscoelastic functionally graded solids. Engineering Analysis with Boundary Elements. 2013; **37**:107-115

[47] Fahmy MA. Generalized magneto-thermo-viscoelastic problems of rotating functionally graded anisotropic plates by the dual reciprocity boundary element method. Journal of Thermal Stresses. 2013;**36**:1-20

[48] Fahmy MA. A three-dimensional generalized magneto-thermo-viscoelastic problem of a rotating functionally graded anisotropic solids with and without energy dissipation. Numerical Heat Transfer, Part A: Applications. 2013;**63**: 713-733

[49] Fahmy MA. Boundary element modeling and simulation of biothermomechanical behavior in anisotropic laser-induced tissue hyperthermia. Engineering Analysis with Boundary Elements. 2019;**101**: 156-164

[50] Fahmy MA. A computerized DRBEM model for generalized magneto-thermo-visco-elastic stress waves in functionally graded anisotropic thin film/substrate structures. Latin American Journal of Solids and Structures. 2014;**11**:386-409

[51] Fahmy MA. A new LRBFCM-GBEM modeling algorithm for general solution of time fractional order dual phase lag bioheat transfer problems in functionally graded tissues. Numerical Heat Transfer, Part A: Applications. 2019;**75**:616-626

[52] Fahmy MA. Design optimization for a simulation of rotating anisotropic viscoelastic porous structures using time-domain OQBEM. Mathematics and Computers in Simulation. 2019;**66**: 193-205

[53] Fahmy MA. Shape design sensitivity and optimization for two-temperature generalized magneto-thermoelastic problems using time-domain DRBEM. Journal of Thermal Stresses. 2018;**41**: 119-138

[54] Fahmy MA. The effect of anisotropy on the structure optimization using golden-section search algorithm based on BEM. Journal of Advances in Mathematics and Computer Science. 2017;**25**:1-18

[55] Fahmy MA. DRBEM sensitivity analysis and shape optimization of rotating magneto-thermo-viscoelastic FGA structures using golden-section search algorithm based on uniform bicubic B-splines. Journal of Advances in Mathematics and Computer Science. 2017;**25**:1-20

[56] Fahmy MA. A predictor-corrector time-stepping DRBEM for shape design sensitivity and optimization of multilayer FGA structures. Transylvanian Review. 2017;**XXV**: 5369-5382

[57] Fahmy MA. Computerized Boundary Element Solutions for Thermoelastic Problems: Applications to Functionally Graded Anisotropic Structures. Saarbrücken, Germany: LAP Lambert Academic Publishing; 2017

[58] Fahmy MA. Modeling and optimization of anisotropic viscoelastic porous structures using CQBEM and moving asymptotes algorithm. Arabian Journal for Science and Engineering. 2019;**44**:1671-1684

[59] Fahmy MA. Boundary Element Computation of Shape Sensitivity and Optimization: Applications to Functionally Graded Anisotropic Structures. Saarbrücken, Germany: LAP Lambert Academic Publishing; 2017

[60] Fahmy MA. Shape design sensitivity and optimization of anisotropic functionally graded smart structures using bicubic B-splines DRBEM. Engineering Analysis with Boundary Elements. 2018;**87**:27-35

[61] Sladek V, Sladek J. Boundary element method in micropolar thermoelasticity. Part I: Boundary integral equations. Engineering Analysis. 1985;**2**:40-50

[62] Sladek V, Sladek J. Boundary element method in micropolar thermoelasticity. Part II: Boundary integro-differential equations. Engineering Analysis. 1985;**2**:81-91

[63] Sladek V, Sladek J. Boundary element method in micropolar thermoelasticity. Part III: Numerical solution. Engineering Analysis. 1985;**2, 2**:155-162

[64] Huang FY, Liang KZ. Boundary element method for micropolar thermoelasticity. Engineering Analysis with Boundary Elements. 1996;**17**:19-26

[65] Goldberg DE. Genetic Algorithms in Search, Optimization and Machine Learning. USA: Addison-Wesley Longman Publishing Co.; 1989

[66] Weile DS, Michielssen E. Genetic algorithm optimization applied to electromagnetics: A review. IEEE Transactions on Antennas and Propagation. 1997;**45**:343-353

[67] Janna C, Ferronato M, Sartoretto F, Gambolati G. FSAIPACK: A software package for high-performance factored sparse approximate inverse preconditioning. ACM Transactions on Mathematical Software. 2015;**41**:10: 1-10:26

[68] Eringen AC. Theory of micropolar elasticity. In: Liebowitz H, editor. Fracture. New York: Academic Press; 1968. pp. 621-729

[69] Dragos L. Fundamental solutions in micropolar elasticity. International Journal of Engineering Science. 1984;**22**: 265-275

[70] Oliveira HL, Leonel ED. Boundary element method applied to topology optimization using the level set method and an alternative velocity regularization. Meccanica. 2019;**54**: 549-563

[71] Itzá R, Viveros UI, Parra JO. Optimal implicit 2-D finite differences to model wave propagation in poroelastic media. Geophysical Journal International. 2016;**206**:1111-1125

Chapter 13

Dynamic Stiffness Method for Vibrations of Ship Structures

Xuewen Yin, Kuikui Zhong, Zitian Wei and Wenwei Wu

Abstract

Initiated by the objective to address the dynamics of ship structures other than conventional finite element method, a dynamic stiffness method (DSM) is proposed systematically including that for three types of element models. A DSM element accounting for both in-plane and bending vibrations in flat rectangular plates is developed, which makes it possible for modeling wave conversion across junctions in built-up plates. In addition, a DSM element for stiffened plates is formulated, which considers all possible vibrations in plates and beams, i.e., bending, torsion, and extension motions. The third type of DSM plate element takes fluid loading into account, which is induced by vibrating plate. Finally, the proposed DSM method is extended to address vibration transmission in a built-up plate structure, which demonstrates the great potentials of DSM in application to more practical and more general engineering fields.

Keywords: dynamic stiffness method, FEM, power flow, beam-stiffened, ship structures

1. Introduction

The vibrational and acoustic characteristics of ship structures are likely to be one of a number of practical concerns not only to mechanical designers and research scientists, but sometimes even to military defense officers. The reasons lie in the following facts. Firstly, excessive vibration levels induced by operating machines or incident waves can inevitably lead to structural fatigue, failure, or even unexpected disasters. Besides, onboard vibration and noise are one of the most important indexes on ship habitability. Too much exposure to such vibrational and noisy environments can make ship crew members uncomfortable, fatigue, or even unhealthy, which has been convinced from a lot of experimental data, and even witnessed from many ship collision accidents. Last but not least, for naval ships, their vibration and acoustic signals make them as attack targets during war time, which also challenge the performances of onboard acoustic instruments.

Up to now, many numerical methods are developed and then utilized in addressing the vibrational and acoustic characteristics of ship structures, which can be found in numerous literatures. Among them are finite element method (FEM), boundary element method (BEM), statistical energy method (SEA), and mesh free methods, etc. Through intense academic efforts from engineers and scientists, and also due to commercial operations from software developer, most them are coded into commercial software, and comprehensively influent the way we design our products almost covering all the engineering fields such as civil engineering, ship and ocean engineering, chemical engineering, and etc. To some extent, we must confess that we, not only engineers, scientists, but product managers, or even government officers, have underestimate the power of novel numerical methods and how much they forge the manufacturing process in modern industries.

As for finite element method, it is one of the most successful numerical methods in high fidelity modeling of the dynamic behaviors of complex structures. To the best of our knowledge, SAP is the first commercial software. Soon after, other software like ADINA, ANSYS, ABAQUS, NASTRAN, and DYNTRAN have been developed and scattered in worldwide universities and industries.

However, like any other numerical methods, FEM has many inherent drawbacks due to the way it discretizes the structures. For instance, to address the vibrational responses in high frequencies, the mesh size must be as tiny as 1/6, or less, of the structural waves so that it can accurately reproduce the dynamics of the structures. However, such a meshing strategy is not always successful since too much finer meshes need not only excessive computational costs, but also lead to unexpected numerical uncertainties.

As for ship structures, the vibration of fluid-loaded plates or shells composes as a very important part in the studies of many engineering structures [1–3]. One of the major reasons lies in the fact that the dynamics of these structures depends on the structures and the fluid simultaneously. The vibrating structures can induce pressure disturbances in their surrounding fluid, and, in return, the resonance frequencies and vibrational responses of the structures can be altered [4, 5].

Recently, dynamic stiffness method (DSM) has won great interests and received intense studies [6–10] from research and design engineers because it can overcome the above issues without too much geometrical discretization requirements. Various DSM elements have been developed for transverse or in-plane vibrations of plates. In the beginning, more research works were mainly focused on transverse vibrations since bending modes are easily excited, especially in low frequencies. Dozens of investigator [6–15] made comprehensive contributions on DSM that only accounts for transverse vibrations of a plate with two opposite edges simply supported. Later, Bercin and Langley [8, 9] proposed a DSM that incorporates both in-plane and bending vibrations. It is reasonably expected that all these works are only applicable to few specified cases due to oversimplified modeling assumptions. To address the vibrations of more practical engineering structures, Casimir et al. [7] developed DSM elements for a plate with completely free boundary conditions, in which Gorman's superposition method was employed to obtain the exact transverse displacements. Banerjee and his colleagues [10–12] proposed the dynamic stiffness matrix for a rectangular plate with arbitrary boundary conditions. Similarly to DSM for bending plates with arbitrary boundary conditions, the dynamic stiffness matrix for in-plane vibrations of plates is developed by Ghorbel et al. [15, 16], Nefovska-Danilovic and Petronijevic [17, 18] in which all the four edges can be prescribed with any arbitrary conditions by adopting Gorman's superposition method.

Since the year 2016, Yin and his associates [19–21] have conducted comprehensive studies on developing dynamic stiffness method and its application to the dynamics of ship structures. Li et al. [19] proposed a dynamic stiffness formulation accounting for both in-plane and bending vibrations of plates with two opposite edges simply supported. This method was then employed for modeling vibration transmission with built-up plate structures [22] and a ship cabin with complex hulls. To consider the dynamics of stiffened plates, Yin et al. [21] extended Li's formulations and developed a dynamic stiffness method that considers torsion, bending, and extension vibrations in beams with eccentric cross-sections.

The main objective of this work is to formulate the vibration analysis of ship structures based on dynamic stiffness method that accounts for both in-plane and bending vibrations within plate itself, all possible motions in stiffened beams, fluid-loading, respectively. The present paper is organized as follows. In Section 2, this dynamic stiffness method is briefly summarized, which present the development of the three types of models. In Section 3, our proposed method is demonstrated by investigating the characteristics of representative plate structures.

2. Development of dynamic stiffness formulations

2.1 Model description

Figure 1 shows multiple rectangular plates in global coordinates OXYZ, which are rigidly joined along their common edges. Each plate has dimension of $L_x \times L_y$ and thickness of h. Its two opposite edges marked by the symbol 'S-S' denote simply supported boundary conditions while the other two edges are arbitrary. In addition, each plate is reinforced by uniform eccentric beams, and in contact with acoustic fluid on its one side.

2.2 Development of plate element

Consider a vibrating flat plate in contact with acoustic fluid on its lower side, which is made of isotropic material with Young's modulus E, bulk density ρ, Poisson's ratio μ, and damping ratio η . Its governing equations for both in-plane and bending vibrations can be written as,

$$\begin{cases} \dfrac{\partial^2 u}{\partial x^2} + a_1 \dfrac{\partial^2 u}{\partial y^2} + a_2 \dfrac{\partial^2 v}{\partial x \partial y} + \dfrac{m\omega^2}{B} u = 0 \\ \dfrac{\partial^2 v}{\partial y^2} + a_1 \dfrac{\partial^2 v}{\partial x^2} + a_2 \dfrac{\partial^2 u}{\partial x \partial y} + \dfrac{m\omega^2}{B} v = 0 \\ D\nabla^4 w - m\omega^2 w = -p_a(x, y, 0) \end{cases} \tag{1}$$

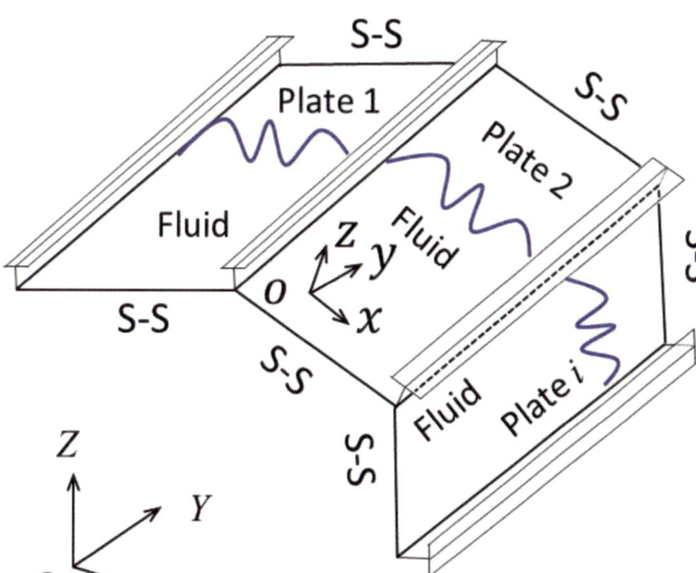

Figure 1.
A built-up plate structure with beam stiffeners.

p_a is the induced acoustic pressure due to the bending vibration of the plate. u, v and w are the displacements in x-, y- and z-directions. m and ω are mass per unit area of the plate and circular frequency, respectively. The parameters a_1 and a_2 in Eq. (1) are defined as

$$a_1 = \frac{1-\mu}{2}, a_2 = \frac{1+\mu}{2} \tag{2}$$

The extension rigidity B and flexural rigidity D can be found in Ref. [19].

According to Bercin and Langley [9], the displacements for the plate, which is simply supported along its two opposite edges, can be expressed as N truncation terms,

$$\begin{cases} u(x,y) = \sum_{n=1}^{N} \left(C_{1n}\lambda_{1n}e^{\lambda_{1n}x} + C_{2n}\lambda_{2n}e^{\lambda_{2n}x} + C_{3n}k_n e^{\lambda_{3n}x} + C_{4n}k_n e^{\lambda_{4n}x}\right)\sin(k_n y) \\ v(x,y) = \sum_{n=1}^{N} \left(C_{1n}k_n e^{\lambda_{1n}x} + C_{2n}k_n e^{\lambda_{2n}x} + C_{3n}\lambda_{3n}e^{\lambda_{3n}x} + C_{4n}\lambda_{4n}e^{\lambda_{4n}x}\right)\cos(k_n y) \\ w(x,y) = \sum_{n=1}^{N} \left(\cos(\alpha_{1n}x)A_{1n} + \sin(\alpha_{1n}x)A_{2n} + \cosh(\alpha_{2n}x)A_{3n} + \sinh(\alpha_{2n}x)A_{4n}\right)\sin(k_n y) \\ \quad if \ \ k^2 \geq {k_n}^2, \end{cases} \tag{3}$$

And, *if* $k^2 < {k_n}^2$, the bending vibrations can be expanded as near-field disturbance,

$$w(x,y) = \sum_{n=1}^{N} \left(\cosh(\alpha_{1n}x)A_{1n} + \sinh(\alpha_{1n}x)A_{2n} + \cosh(\alpha_{2n}x)A_{3n} + \sinh(\alpha_{2n}x)A_{4n}\right)\sin(k_n y) \tag{4}$$

where $k^2 = \sqrt{\rho\omega^2/D}$ and $k_n = n\pi/L_y$. $C_{mn}, m = 1,2,3,4$ and $A_{mn}, m = 1,2,3,4$ are the unknown constants. Wavenumbers for in-plane and out-of-plane waves take the following forms:

$$\begin{cases} \lambda_{1n,2n} = \pm\sqrt{{k_n}^2 - {k_L}^2}, \lambda_{3n,4n} = \pm\sqrt{{k_n}^2 - {k_T}^2} \\ k^2 \geq {k_n}^2, \alpha_{1n} = \sqrt{k^2 - {k_n}^2}, \alpha_{2n} = \sqrt{k^2 + {k_n}^2} \\ k^2 < {k_n}^2, \alpha_{1n} = \sqrt{{k_n}^2 - k^2}, \alpha_{2n} = \sqrt{{k_n}^2 + k^2} \end{cases} \tag{5}$$

where ${k_L}^2 = \rho\omega^2(1-\mu^2)/E$, ${k_T}^2 = 2\rho\omega^2(1+\mu)/E$.

Accordingly, the transverse shear force Q_x perpendicular to xy plane, the bending moment M_{xx}, longitudinal force N_{xx}, and in-plane shear force N_{xy} along the plate junctions can be derived as follows,

$$\begin{cases} Q_x = -D\left(\frac{\partial^3 w}{\partial x^3} + (2-\mu)\frac{\partial^3 w}{\partial x \partial y^2}\right) \\ M_{xx} = -D\left(\frac{\partial^2 w}{\partial x^2} + \mu\frac{\partial^2 w}{\partial y^2}\right) \\ N_{xx} = -B\left(\frac{\partial u}{\partial x} + \mu\frac{\partial v}{\partial y}\right) \\ N_{xy} = -Ba_1\left(\frac{\partial u}{\partial y} + \frac{\partial v}{\partial x}\right) \end{cases} \tag{6}$$

Based on Eqs. (3) and (6), for any nth mode, the generalized displacement vector $\overline{q}_n$ and force vector $\overline{Q}_n$ are written as,

$$\overline{q}_n = \begin{Bmatrix} \overline{u}_n|_{x=0} \\ \overline{v}_n|_{x=0} \\ \overline{w}|_{x=0} \\ \overline{\theta}_n|_{x=0} \\ \overline{u}_n|_{x=Lx} \\ \overline{v}_n|_{x=Lx} \\ \overline{w}_n|_{x=Lx} \\ \overline{\theta}_n|_{x=Lx} \end{Bmatrix} = \begin{Bmatrix} \int_0^{L_y} \frac{2}{L_y} u(0,y)\sin(k_n y)dy \\ \int_0^{L_y} \frac{2}{L_y} v(0,y)\cos(k_n y)dy \\ \int_0^{L_y} \frac{2}{L_y} w(0,y)\sin(k_n y)dy \\ \int_0^{L_y} \frac{2}{L_y} \theta(0,y)\sin(k_n y)dy \\ \int_0^{L_y} \frac{2}{L_y} u(L_x,y)\sin(k_n y)dy \\ \int_0^{L_y} \frac{2}{L_y} v(L_x,y)\cos(k_n y)dy \\ \int_0^{L_y} \frac{2}{L_y} w(L_x,y)\sin(k_n y)dy \\ \int_0^{L_y} \frac{2}{L_y} \theta(L_x,y)\sin(k_n y)dy \end{Bmatrix}_{8\times 1}, \tag{7}$$

$$\overline{Q}_n = \begin{Bmatrix} \overline{N}_n|_{x=0} \\ \overline{T}_n|_{x=0} \\ \overline{S}_n|_{x=0} \\ \overline{M}_n|_{x=0} \\ \overline{N}_n|_{x=Lx} \\ \overline{T}_n|_{x=Lx} \\ \overline{S}_n|_{x=Lx} \\ \overline{M}_n|_{x=Lx} \end{Bmatrix} = \begin{Bmatrix} \int_0^{L_y} \frac{2}{L_y} N(0,y)\sin(k_n y)dy \\ \int_0^{L_y} \frac{2}{L_y} T(0,y)\cos(k_n y)dy \\ \int_0^{L_y} \frac{2}{L_y} S(0,y)\sin(k_n y)dy \\ \int_0^{L_y} \frac{2}{L_y} M(0,y)\sin(k_n y)dy \\ -\int_0^{L_y} \frac{2}{L_y} N(L_x,y)\sin(k_n y)dy \\ -\int_0^{L_y} \frac{2}{L_y} T(L_x,y)\cos(k_n y)dy \\ -\int_0^{L_y} \frac{2}{L_y} S(L_x,y)\sin(k_n y)dy \\ -\int_0^{L_y} \frac{2}{L_y} M(L_x,y)\sin(k_n y)dy \end{Bmatrix}_{8\times 1} \tag{8}$$

Hence, the relationship between generalized displacements $\overline{q}_n$ and generalized forces $\overline{Q}_n$ at any n_{th} mode can be developed after simple matrix algorithm, which is generally known as dynamic stiffness matrix $\overline{K}_n$. Once the dynamic stiffness matrix is obtained, the dynamic responses resulted from excitations can be readily achieved after solving linear equations like those in conventional finite element methods [19].

2.3 Development of beam element

As shown in **Figure 2**, a beam with an eccentric cross section is located with geometric center O and the shear center G. Based on classical beam theory, the governing equations for the forced vibrations at line $G - G'$ are expressed as,

$$\begin{cases} \dfrac{\partial^2}{\partial y^2}\left(E_r I_z \dfrac{\partial^2 u_r}{\partial y^2}\right) - m_r\omega^2 u_r + m_r\omega^2 z_G\phi_r = P_r \\ E_r \dfrac{\partial^2 v_r}{\partial y^2} + \rho_r\omega^2 v_r = N_r \\ \dfrac{\partial^2}{\partial y^2}\left(E_r I_x \dfrac{\partial^2 w_r}{\partial y^2}\right) - m_r\omega^2 w_r - m_r\omega^2 x_G\phi_r = Q_r \\ G I_t \dfrac{\partial^2 \phi_r}{\partial y^2} + I_0\omega^2\phi_r - m_r\omega^2 x_G w_r + m_r\omega^2 z_G u_r = T_r \end{cases} \tag{9}$$

where u_r, v_r and w_r are the displacements in x_r-, y_r- and z_r-directions, and ϕ_r is the rotation about y_r axis. P_r, N_r, Q_r are the forces acting line $G - G'$ in x_r-, y_r- and z_r-directions, and T_r is the torsion moment about y_r axis. I_x and I_z are the principle moments of the beam's cross-section about x_r- and z_r-axes. E_r and ρ_r are Young's modulus and density of the material. m_r is mass per unit length of the beam, i.e., $\rho_r A_r$, where A_r is the cross-sectional area. G and I_0 are shear modulus of the material, polar moment of mass inertia with respect to shear center, respectively, and I_t is cross-sectional factor in torsion.

Since the beam is attached to one edge of the plate, its motions are in the similar forms as that expressed in Eq. (3) and can be readily written as,

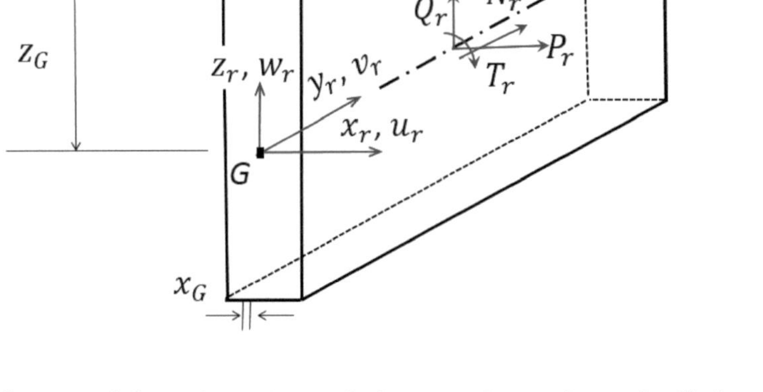

Figure 2.
Schematic illustration of a beam: Geometric center G*, shear center* O*;* x_G *and* z_G *are the offset between* G *and* O *in* x_r*-, and* z_r*-directions, respectively.*

$$\begin{cases} u_r(y_r) = \sum_{n=1}^{N} u_{rn} \sin(k_n y_r) \\ v_r(y_r) = \sum_{n=1}^{N} v_{rn} \cos(k_n y_r) \\ w_r(y_r) = \sum_{n=1}^{N} w_{rn} \sin(k_n y_r) \\ \phi_r(y_r) = \sum_{n=1}^{N} \phi_{rn} \sin(k_n y_r) \end{cases} \tag{10}$$

Substituting Eq. (10) into Eq. (9), and utilizing the orthogonality relationship of the modes, the vibration motions at the n_{th} mode for the beam can be readily derived,

$$\begin{cases} (E_r I_z k_n^4 - m_r\omega^2)u_{rn} + m_r\omega^2 z_G \phi_{rn} = P_{rn} \\ (-E_r A k_n^2 + m_r\omega^2)v_{rn} = N_{rn} \\ (E_r I_x k_n^4 - m_r\omega^2)w_{rn} - m_r\omega^2 x_G \phi_{rn} = Q_{rn} \\ (-GI_t k_n^2 + I_0\omega^2)\phi_{rn} - m_r\omega^2 x_G w_{rn} + m_r\omega^2 z_G u_{rn} = T_{rn} \end{cases} \tag{11}$$

Without complex derivation procedure, Eq. (11) can be rewritten in a more compact matrix form,

$$\overline{F}_{rn} = \overline{K}_{rn}(\omega)\overline{q}_{rn}, \tag{12}$$

where the dynamic stiffness matrix has the following expressions:

$$\overline{K}_{rn} = \begin{bmatrix} E_r I_z k_n^4 - m_r\omega^2 & 0 & 0 & m_r\omega^2 z_G \\ 0 & -E_r A k_n^2 + m_r\omega^2 & & 0 \\ 0 & 0 & E_r I_x k_n^4 - m_r\omega^2 & -m_r\omega^2 x_G \\ m_r\omega^2 z_G & 0 & -m_r\omega^2 x_G & -GI_t k_n^2 + I_0\omega^2 \end{bmatrix}. \tag{13}$$

2.4 Development of fluid-loaded element: acoustic pressure

The acoustic pressure satisfies the Helmholtz equation,

$$\left(\frac{\partial^2}{\partial x^2} + \frac{\partial^2}{\partial y^2} + \frac{\partial^2}{\partial z^2}\right)p_a + k_0^2 p_a = 0, \tag{14}$$

where k_0 is the acoustic wavenumber. The boundary condition at the interface between the plate and the fluid is expressed as

$$\left.\left(\frac{\partial p_a}{\partial z}\right)\right|_{z=0} = \rho_0\omega^2 w, \tag{15}$$

where ρ_0 is the density of the acoustic fluid. Since the acoustic pressure p_a has the following form:

$$p_a(x, y, z) = |p_a| \exp\left[-j(k_x x + k_y y + k_z z)\right], \tag{16}$$

where $|p_a|$ is the amplitude of the acoustic pressure, and k_x, k_y, and k_z are wavenumbers for the acoustic waves. It is ready to obtain the expression for the acoustic pressure at the plate-fluid interface,

$$p_a(x,y,0) = \begin{cases} \dfrac{j\rho_0\omega^2 w}{\left(k_0^2 - k_b^2\right)^{½}}, & if \quad k_b < k_0 , \\ \dfrac{-\rho_0\omega^2 w}{\left(k_b^2 - k_0^2\right)^{½}}, & if \quad k_b > k_0 . \end{cases} \tag{17}$$

It is noted that we have the expression $k_b^2 = k_x^2 + k_y^2$, where k_b is the wavenumbers for the structural waves propagating within the plates. For sake of brevity, the relationship between the acoustic pressure at the fluid–structure interface and the inertia terms due to the vibration of the plate, which is referred to as fluid-loading parameter, can be rewritten as,

$$\varepsilon_f = p_a(x,y,0) \big/ \left[m\omega^2 w\right] = \begin{cases} \dfrac{j\rho_0}{m\left(k_0^2 - k_b^2\right)^{½}}, & if \quad k_b < k_0 , \\ \dfrac{-\rho_0}{m\left(k_b^2 - k_0^2\right)^{½}}, & if \quad k_b > k_0 . \end{cases} \tag{18}$$

2.5 Dynamic responses of built-up plate structures

The dynamic stiffness matrices for the plate and the beam (in Sections 2.2 and 2.3) are expressed in local coordinates, which can be termed as local dynamic stiffness matrices. With reference to the conventional finite element technique, the dynamic stiffness matrix for each plate element and each beam element can be readily assembled into overall global dynamic stiffness matrix. Hence, the dynamic responses of a built-up structure composed of plates and beams can be solved through novel numerical methods.

3. Numerical results and discussion

Without loss of generality, we only focus on the vibration transmission in a built-up plate structure that is reinforced by stiffeners or plates. Numerical results for the dynamics of plates with beam stiffeners based on our method can found in [21].

3.1 Transmission modes within a plate stiffened by stiffeners

To demonstrate our method in addressing the vibration transmission within complex built-up structures, a horizontal plate reinforced by a vertical plate, i.e., plate 2 is employed in this subsection. The detailed parameters of the plates are listed in **Table 1**. The two opposite long edges of plate 1 is simply supported. One of the free end of the plate, namely, left edge, is subjected to uniformly distributed vertical forces of 1 N/m.

Yin et al. [22] identify that there are three representative transmission modes in a stiffened plate. As the plate structures get more complex, similar phenomena can be also found, in which a plate is stiffened by 9 identical plates. When the left side of the plate is enforced with transverse force, three representative transmission

	E (Gpa)	ρ(Kg/m^3)	μ	η	L_x(m)	L_y(m)	h(m)
Plate 1	200	7800	0.3	0.01	6.0	1.0	0.008
Plate 2	200	7800	0.3	0.01	0.5	1.0	0.008

No beam stiffeners are considered in case 1.

Table 1.
Geometry and material parameters of the plates (case 1).

modes can be clearly identified. In **Figure 3(a)**, only the left local portion of the plates is excited that implies bending waves cannot propagate effectively forward due the presence of the stiffening plates. However, in some frequency regimes as shown in **Figure 3(b)** and **(c)**, bending waves can pass the stiffening members freely. As frequency increases, the stiffening members act more like a barrier that prevent structural waves propagate.

From **Figure 3(a)–(d)**, we can convince that the vibration transmission modes do exist in even more complex plate structures. In addition, we suggest to explore the underlying mechanisms, if any, between these transmission modes and the well-known pass band and stop band since vibration transmission is probably one of the most important characteristics in complex plate structures, e.g., ship structures, etc.

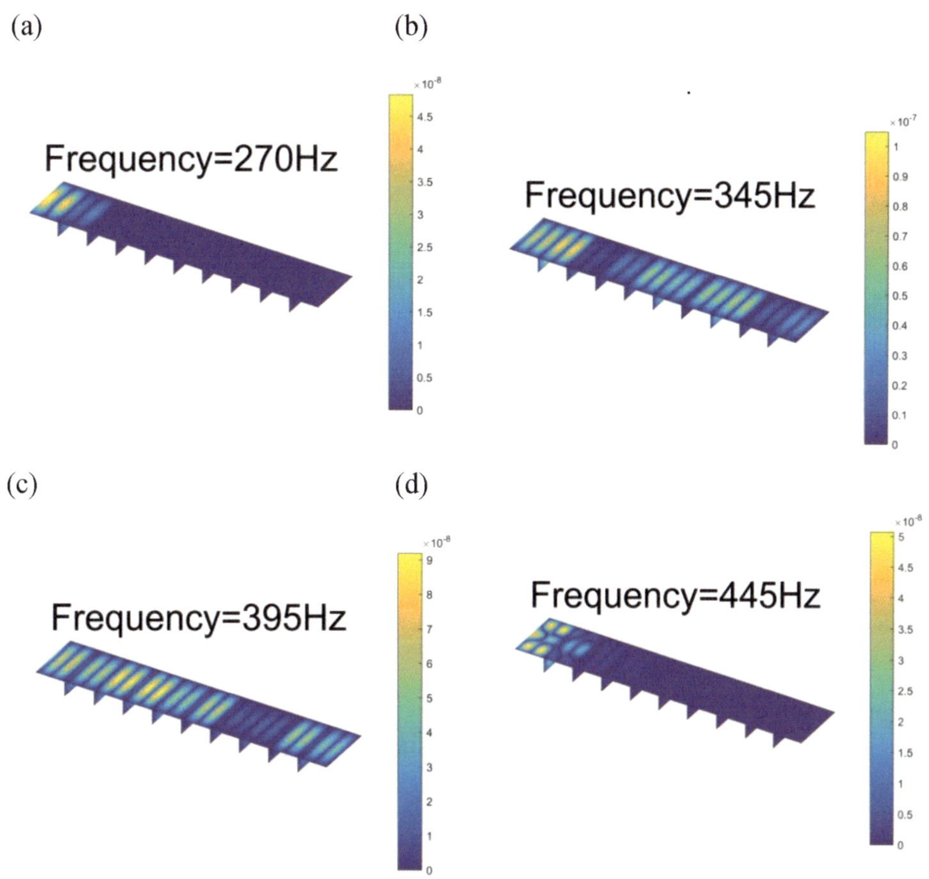

Figure 3.
Representative vibrational transmission modes of a stiffened plate: (a) 270 Hz, (b) 345 Hz, (c) 395 Hz, and (d) 445 Hz.

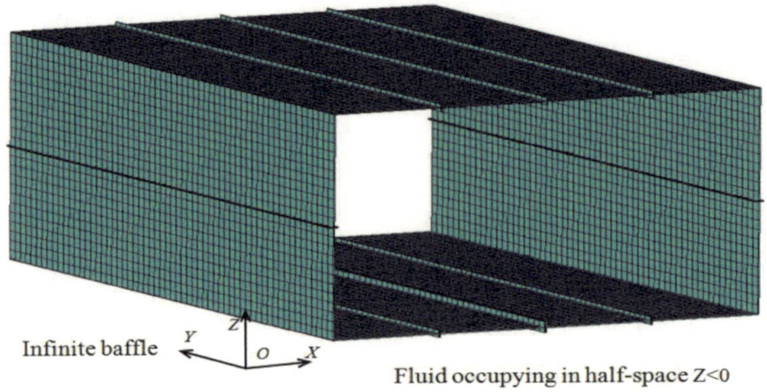

Figure 4.
A ship hull reinforced with eight stiffeners.

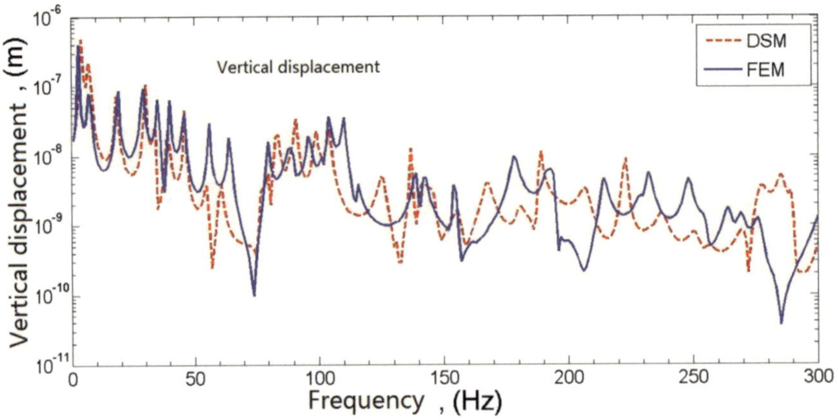

Figure 5.
Vertical displacement at the middle point in the bottom plate.

3.2 Vibrations of a ship hull in contact with water

Figure 4 shows a ship hull that is reinforced by eight beams with dimension $0.02m \times 0.02m$ along the junctions of their neural planes. The ship hull has the dimension of $6m \times 4m \times 2.4m$ and with thickness of 0.008 m. The bottom of the ship hull is in contact with water. About 1 N concentrated force is applied at the middle point in upper plate and the response gauge is set at middle point in the bottom plate.

Figure 5 shows the curves for the vertical displacement obtained by FEM and DSM, respectively. The truncation term N is set to 6 in DSM and the mesh size in the FEM is $0.2m \times 0.2m$. It is indicated that satisfactory agreement can be found between the results from DSM and those from FEM, which implies that our proposed method can provide excellent numerical results for ship structures.

4. Conclusion

A DSM is proposed to address the dynamics of ship structures, which include three types of elements. First, a DSM formulation for both in-plane and bending vibrations in flat rectangular plates is developed. Then, a DSM for stiffening beams is addressed, which accounts for all possible vibrations in plates and beams, i.e., bending, torsion, and extension motions. Finally, a DS plate element with fluid loadings included is formulated. The numerical results for the vibrations for a ship hull based on the proposed DSM have excellent agreement with those results obtained from FEM, which demonstrate its potential in addressing the dynamics of ship structures. In addition, vibration transmission modes of a stiffened plate are also addressed using this method.

Acknowledgements

This work was partially supported by High-Tech Ship Fund from the Ministry of Industry and Information Technology (MIIT): Deepwater Semi-submersible Support Platform (No.: 2016 [546]), High Quality Brand Ship Board Machinery (No.: 2016 [547]). The authors would like to thank Provincial Youth Fund (No.: BK20170217) from Jiangsu Science Foundation.

Author details

Xuewen Yin*, Kuikui Zhong, Zitian Wei and Wenwei Wu
National Key Laboratory on Ship Vibration and Noise, China Ship Scientific Research Center, Wuxi, China

*Address all correspondence to: x.w.yin@cssrc.com.cn

References

[1] Mace BR. Sound radiation form fluid-loaded orthogonally stiffened plates. Journal of Sound and Vibration. 1981; **79**(3):439-452

[2] Mead DJ. Plates with regular stiffening in acoustic media: Vibration and radiation. Journal of the Acoustical Society of America. 1990;**88**(1):391-401

[3] Missaoui J, Cheng L, Richard MJ. Free and forced vibration of a cylindrical shell with a floor partition. Journal of Sound and Vibration. 1996; **190**(1):21-40

[4] Yin XW, Gu XJ, Cui HF, et al. Acoustic radiation from a laminated composite plate reinforced by doubly periodic parallel stiffeners. Journal of Sound and Vibration. 2007;**306**:877-889

[5] Yin XW, Liu LJ, Hua HX, Shen RY. Acoustic radiation from an infinite laminated composite cylindrical shell with doubly periodic rings. ASME Journal of Vibration and Acoustics. 2009;**131**(1):011005

[6] Kevorkian S, Pascal M. An accurate method for free vibration analysis of structures with application to plates. Journal of Sound and Vibration. 2001; **246**(5):795-814

[7] Casimir JB, Kevorkian K, Vinh T. The dynamic stiffness matrix of two-dimensional elements: Application to Kirchhoff's plate continuous elements. Journal of Sound and Vibration. 2005; **287**:571-589

[8] Langlegy RS. Application of the dynamic stiffness method to the free and forced vibrations of aircraft panels. Journal of Sound and Vibration. 1989; **135**(2):319-331

[9] Bercin AN, Langley RS. Application of the dynamic stiffness technique to the in-plane vibrations of plate structures. Computers and Structures. 1996;**59**(5):869-875

[10] Boscolo M, Banerjee JR. Dynamic stiffness method for exact inplane free vibration analysis of plates and plate assemblies. Journal of Sound and Vibration. 2011;**330**:2928-2936

[11] Boscolo M, Banerjee JR. Dynamic stiffness formulation for composite Mindlin plates for exact modal analysis of structures. Part I: Theory. Computers and Structures. 2012;**96**:61-73

[12] Boscolo M, Banerjee JR. Layer-wise dynamic stiffness solution for free vibration analysis of laminated composite plates. Journal of Sound and Vibration. 2014;**333**:200-227

[13] Danial AN, Doyle JF, Rizzi SA. Dynamic analysis of folded plate structures. Journal of Vibration and Acoustics. 1996;**118**:591-598

[14] Fazzolari FA. A refined dynamic stiffness element for free vibration analysis of cross-ply laminated composite cylindrical and spherical shallow shells. Composites Part B: Engineering. 2014;**62**:143-158

[15] Ghorbel O, Casimir JB, Hammami L, et al. Dynamic stiffness formulation for free orthotropic plates. Journal of Sound and Vibration. 2015;**346**:361-375

[16] Ghorbel O, Casimir JB, Hammami L, et al. In-plane dynamic stiffness matrix for a free orthotropic plate. Journal of Sound and Vibration. 2016;**364**:234-246

[17] Kolarevic N, Nefovska-Danilovic N, Petronijevic M. Dynamic stiffness elements for free vibration analysis of rectangular Mindlin plate assemblies. Journal of Sound and Vibration. 2015; **359**:84-106

[18] Nefovska-Danilovic M, Petronijevic M. In-plane free vibration and response analysis of isotropic rectangular plates using the dynamic stiffness method. Computers and Structures. 2015;**152**:82-95

[19] Li H, Yin XW, Wu WW. Dynamic stiffness formulation for in-plane and bending vibrations of plates with two opposite edges simply supported. Journal of Vibration and Control. 2018; **24**(9):1652-1669

[20] Wu WW, Li H, Yin XW. Power flow analysis of built-up plate structures using the dynamic stiffness method. Journal of Vibration and Control. 2018; **24**(13):2815-2831

[21] Yin XW, Wu WW, Li H. Dynamic stiffness formulation for the vibrations of stiffened plates with consideration of in-plane deformations. Journal of Vibration and Control. 2018;**24**(20): 4825-4838

[22] Yin X, Wu W, Li H. Vibration transmission within beam-stiffened plate structures using dynamic stiffness method. Procedia Engineering. 2018; **199**:411-416